From Ancient Time

to the Present

From Ancient Times to the Present

The major economic ideas and doctrines from ancient days to the present, the vital role of economics in everyday life, the causes of depressions, inflation, and unemployment—all these and many other important economic issues are discussed in this compact and instructive volume by one of today's leading authorities.

Clearly and simply written by the well-known expert, George Soule, this provocative book examines the views, times, and lives of such outstanding economic thinkers as Comte, Malthus, John Stuart Mill, Thorstein Veblen, and many others. Mr. Soule traces the development of modern economic theories from the strict governmental control of Mercantilism to Adam Smith's arguments for *laissez faire* and freedom of private enterprise, from Karl Marx's denunciations of the capitalistic order to George Maynard Keynes's support of government investment to prevent depression.

Thought-provoking and timely, this remarkably concise book effectively clarifies for the layman the vast and important field of economics which underlies the whole structure of society.

THIS IS A REPRINT OF THE ORIGINAL HARDCOVER EDITION
PUBLISHED BY THE VIKING PRESS, INC.

GEORGE SOULE

IDEAS of the GREAT ECONOMISTS

A MENTOR BOOK
Published by THE NEW AMERICAN LIBRARY

To my daughter Marcia

*Published as a MENTOR BOOK
by arrangement with The Viking Press, Inc.,
who have authorized this softcover edition.*

SEVENTH PRINTING

MENTOR TRADEMARK REG. U.S. PAT. OFF. AND FOREIGN COUNTRIES
REGISTERED TRADEMARK—MARCA REGISTRADA
HECHO EN CHICAGO, U.S.A.

MENTOR BOOKS are published *in the United States* by
The New American Library of World Literature, Inc.,
501 Madison Avenue, New York, New York 10022.
in Canada by The New American Library of Canada Limited,
156 Front Street West, Toronto 1, Ontario,
in the United Kingdom by The New English Library Limited,
Barnard's Inn, Holborn, London, E.C. 1, England

PRINTED IN THE UNITED STATES OF AMERICA

Contents

1

What Economics Is

Economic practices appeared in real life long before any theory about them existed. They were not parts of a system thought out as a whole and deliberately installed, but an aggregation of particular ways of meeting practical needs, invented by nobody knows whom. They were tried, modified, combined, retained, or abandoned according to change and circumstance.

Money, prices, markets, profit, interest, wages, investment, taxes and other familiar economic terms are names for habits or institutions found in many human communities. They are part of the rich and varied texture of experience, just as are other things not commonly thought of as economic, like dress, tools, weapons, art, education, marriage, government, or religion.

The kind of behavior usually regarded as economic is that most closely connected with the ways in which men make their livings. How does a family, a tribe, a nation, a group of nations, produce and distribute the food, clothing, shelter, services, and other things that people desire? How does it accumulate—or dissipate—material wealth? Such practices vary widely among the many peoples who now inhabit the earth. Economic habits have changed markedly over the centuries even in the type of culture with which we are most familiar—the European, with its offshoots in America and other continents.

Housekeeping by Family and State

Though no systematic study of economic as distinguished from other aspects of human behavior was made until modern times, the word is derived from one in the language of the ancient Greeks—*oikonomos*. This word means, as far as a translation is possible, household. The housekeeper has to see that there is enough food, clothing, and shelter, that the house is kept in order, that the necessary duties are performed by appropriate members of the household, and that their products are distributed according to necessity or custom.

A household prospers not automatically, nor solely by gifts of a benevolent nature, but according to the skill of the housekeeper and the diligence and deftness of its members. Thus, economic thinking immemorially has included judgments as to better or worse ways of doing things, approval or disapproval of methods, a sense of order and efficiency. It implies management in the interest of a community, small or larger; it implies choices of policy and program. Economic opinions have almost always been associated with moral codes about the manner of work and the division of the product.

As the ancient Greeks used the term, it applied not merely to literal households, but to the city-state, which was the characteristic form of Greek government. The name of that state being *polis*, "political economy" has been used as the name of the subject, even until recent times. It is a good and useful name at present, since one can scarcely conceive of a modern economic society in which government does not, by necessity, play a large role. Men talk increasingly about the economies of particular states, like the American economy, the British economy. Perhaps it would clear up confusion if the Greek words were abandoned altogether, and instead of teaching "economics" in our schools we named the subject "national (or international) housekeeping."

Ancient Economic Ideas

Most records of economic ideas in ancient times are embodied in religious teachings, codes of law, or moral exhortations. Archaeologists have discovered in the Babylonian code of Hammurabi detailed regulations for economic practices. The Bible, reflecting the theocratic state of the ancient Jews, contains many injunctions against greed and extortion, and against overemphasis on material wealth. Justice and mercy are enjoined on the people in their economic relationships. These admonitions of the prophets bear witness that in the evolution of the Hebrews from the life of a primitive tribe to a more highly commercial order, many fissures and strains developed in their society.

Greek society experienced a similar evolution. Tribal at the beginning, it consisted of largely self-sustaining households. Kings and priests ruled. Gradually a landed aristocracy came into being, while peasants and artisans lacked any important stake in the soil. Wars, with their captives, created a separate slave class to do manual labor. As growing navigation and commerce yielded wealth to merchants, businessmen found themselves in conflict with the ruling hereditary landlords. Because of the strains of a changing society, lawgivers, politicians, and philosophers tried to develop principles for the regulation of economic relationships, as well as for other kinds of behavior and the structure of the state. Democracy as eventually practiced in Athens was in large part an expression of the interests of commerce. Of course it excluded the slaves and the other workers in handicrafts.

Plato, who lived during the maturity of Athenian culture in the fourth century B. C., shared the aristocratic tradition, which looked down upon both manual labor and the pursuit of wealth but exalted the warrior, the statesman, and those responsible for agriculture. He showed aversion to the emphasis on gain-seeking which had accompanied the rise of commerce. In *The Republic* he outlined what he regarded as an ideal state, in which the rulers were to be educated for their responsibility from childhood and chosen by competitive examination. Craftsmen were to be excluded from political rights since their occupations would prevent them from devoting the necessary attention to the high duties of citizenship. This division implied no similar gulf in riches, since the ruling class was to possess no property beyond that necessary to support them. Common ownership of property was to be the general rule.

Plato recognized that production is the basis of the state and observed that a diversity of occupations made a community necessary, since "no one is self-sufficing, but all of us have many wants. . . . All things are produced more plentifully and easily, and of a better quality, when one man does one thing which is natural to him and does it at the right time, and leaves other things." The doctrine of specialization was later carried much further by Adam Smith.

Too large a state was distasteful to Plato. It must be large enough to provide opportunity to varying talents, but not so large that the citizens could not know one another, or that it would be clumsy to administer. Too much catering to demand for luxuries would necessitate too large a community and stimulate seeking of gain for its own sake. Plato fixed on

5040 as the best number of lots or establishments—curiously enough, because 5040 is divisible by all numbers up to and including ten.

Later Classical Ideas

Aristotle was more an observer than his predecessor Plato; his generalizations foreshadow those of the modern scientist since they depend more upon study of the data available to him.

Aristotle explicitly based his economic opinions on good management of the household. The necessaries of life, he maintained, constitute true wealth; but nothing is wealth if it is in excess of the need for it or is not made for use. He approved "the natural art of acquisition which is practiced by managers of households and statesmen," but not commerce carried on for gain. Exchange of commodities he sanctioned; "it arises at first in a natural manner from the circumstances that some have too little, others too much,"—of any given article. But retail trade "is not a natural part of the art of money-making; had it been so, men would have ceased to exchange when they had enough."

On this ground Aristotle reserved his hearty disapproval for currency and the lending of money at interest. Money may be a useful instrument of exchange, but, thought Aristotle, when it tempts people to pile up unused gains or accumulate wealth by lending money, it is "sterile" or unproductive, and it promotes disparity in riches and financial irregularities.

Aristotle, like Plato, placed a high value on management of agriculture but thought that workers in industry should not be citizens. The "vulgar" arts "utterly ruin the bodies of workers and managers alike, compelling men, as they do, to lead sedentary lives and huddle indoors, or in some cases to spend the day before a fire. Then as men's bodies become enervated, so their souls grow sicklier. And these vulgar crafts involve complete absence of leisure and hinder men from social and civic life; consequently men such as these are bad friends and indifferent defenders of their country." Unlike Plato, Aristotle thought that common ownership of property was impracticable and contrary to the dispositions of men, hindering both pride of ownership and generous impulses.

Xenophon, the historian and soldier, like Plato a disciple of Socrates, actually wrote a treatise called *Oikonomikos*—household management. He lauded agriculture as the basis of economic wealth, favored encouragement of shipping and trade by the state. More than either Plato or Aristotle, his writings suggest some of the aspects of modern capitalism. He advocated an increase of silver mining to add to the general wealth and encourage trade; he favored joint-stock companies in which individuals would combine to carry on business; he regarded the arts of peace as more rewarding than those of war; and he sanctioned large cities because they would favor a higher degree of specialization and division of labor. Yet he approved of slavery and proposed public ownership of silver mining and other activities.

Roman economic thought was scanty and in general was based on the Greek. Like the Greek and in some measure the Hebrew too, it reflected values derived from the tradition of an agrarian and military society and was a reaction against what was regarded as the corruption introduced by commercialism, against luxury bred by love of money for its own sake and by undue disparities in wealth.

Medieval Economic Thinking

As the Roman Empire disintegrated and the feudal order became established in Europe, economic relationships as usual adjusted themselves to the structure of society. Landed aristocracy was in the saddle, agriculture was the principal occupation, most production was carried on by dependents of the feudal lords. Within the domain of each lord the distribution of the product was regulated not by purchase and sale but by traditional rules of sharing. The lords themselves were graduated in a hierarchy. Society was stratified in classes, each of which had a fixed status, with certain duties and rights.

Though markets and the use of money covered far less of the economy than at present, there were traders, moneychangers, and independent handicraftsmen, with their apprentices. The various merchants and crafts became organized in guilds—somewhat like the trade associations of businessmen today. These guilds established standards of skill, the prices of purchase and sale, the rates of wages.

The Christian church, by this time nearly universal in Europe, vied with the temporal authorities for power and strove to impose a moral order on the institutions of the time. As before, economic doctrines arose largely from this effort and were developed not as a separate body of theory but as part of a general moral code.

Virtually all the scholars and writers were churchmen. They, in turn, regarded Aristotle as the great authority in scientific and mundane matters, including ethics. There were, of course, frequent controversies in actual economic relationships about the sharing of the product, the prices charged by craftsmen, and the wages paid (when money wages were paid at all). Such controversies the church attempted to settle by ideas of fairness rather than by sanctioning the action of demand and supply in a market.

St. Thomas Aquinas, who dominated the thinking of the period, took over Aristotle's conception that justice could be divided into two categories: distributive justice, applying to the distribution of the product of the household, feudal estate, or other economic entity; and compensatory justice, applying to the exchange of goods or services. In distributive justice, income should be that which was customary; it should be suitable to the station of the recipient. In exchange, prices should compensate both parties fairly for the products they had to offer. "Wherever a good is to be found, its essence consists in its due measure." The just wage and the fair price are perennial ideas in economic doctrine.

Aquinas, like the other church authorities, condemned the charging of interest on borrowed money, or, as interest was called, "usury," because, as Aristotle thought, money is properly only a medium of exchange and does not itself produce anything. But to this general doctrine Aquinas made exceptions in cases when the money did not actually change hands but was simply withheld for a period, as in the payment of rent or hire or payment for goods bought. Later refinements of the theory sanctioned the payment of interest for missed opportunities of gain on the part of the lender (*lucrum cessans*), loss incurred by or injury to the lender (*damnum emergens*), risk of loss through possible non-payment, and delay in repayment after the agreed time.

Emphasis on the iniquity of usury was not merely an abstract idea; like most other prominent economic doctrines then and now, it served a purpose important at the time to those who promulgated it. As in the approaching maturity of ancient Greek states, where the doctrine originated, trade and markets had a tendency to grow, the use of money was increasing; successful merchants and money-changers accumulated money. There was demand for money on the part of the old ruling classes, so that those who had more than enough of it for their own needs could readily charge interest on loans.

But the growth of capital accumulation by lending not only occasionally wiped out individual borrowers who believed they were victims of monopolists of money: it seemed to threaten the structure of a society producing not for gain but for use, in which each participant was supposed to receive his just share. The medieval church and its feudal allies correctly sensed a menace to their security and their power in the growth of capitalism—though nobody yet had called it that. Condemnation of usury was a symptom that the feudal order was being undermined by new ways of carrying on production and exchange.

Essence of Early Economic Doctrines

The recorded economic ideas of the ancient world, and of the medieval European order which looked back to it for intellectual guidance, again and again emphasized themes which frequently have reappeared in more modern thought. Among these were:

The primacy of agriculture as the basis of all other means of life.

The concept that economic practices and orders were closely related to some social unit in which management could be exercised, like a household, a city-state, a feudal domain, an organized group of traders or craftsmen.

The desirability of wise and prudent management of economic processes.

The legitimacy of production and exchange so long—and only so long—as the product was destined for use.

The Biblical warning that "love of money is the root of all evil."

The feeling that certain types of occupation are ignoble and unfit men to participate in public life or government. Among these were manual labor without leisure or opportunity for education, and pursuit of gain for its own sake.

The belief that distribution and exchange of goods should be regulated justly and that this end could be achieved by a society in which each contributed what he was best fitted or destined to do and received in return a "fair" or customary share of the general product.

Such doctrines were not a subject of study kept separate in a water-tight compartment from other observations of human behavior. They were parts of general codes of morals or philosophy that aimed to cover the whole range of human experience, both material and spiritual.

Most economic doctrines were occasioned by disturbing changes in human society, which brought disorder and discomfort to many and challenges to tradition. The doctrines always constituted an effort to effectuate some social purpose. Often they were, in one aspect, voiced in defense of a declining priesthood or landed aristocracy. Though the record does not contain many writings justifying those who wanted change for the benefit of the lower classes or the innovators, such movements occurred from time to time, and doubtless would have left more records if their leaders had been more literate.

What Economics Is

From the record of ancient and medieval economic writings a justified inference is that economics was, at least in those days:

Not the actual economic practices in force at any one time but rather ideas or doctrines prompted by those practices;

Not a system of "natural laws" demonstrable by experiment, making possible prediction, and good for any age and locality, like the laws of physics or chemistry, since economic doctrines varied with varying customs and stages of development;

Not a wholly objective and disinterested description of economic phenomena, since all bodies of economic doctrine had a purpose other than accurate explanation or analysis— a moral purpose, a civic purpose, or the defense of a past culture, an existing culture, or one proposed for the future;

Not a body of learning which could be separated from study of other aspects of man's behavior, such as his political, social, aesthetic, religious, or ethical life.

The modern tradition, beginning with the Renaissance, the Reformation, and the eighteenth-century "enlightenment," made a break with ancient and medieval thought. It introduced new concepts of science and scientific method, ideas of individual liberty, freedom of conscience, social and political equality, the substitution of contract based on choice for class status, a humanism expressed in hope for better things in this world rather than merely salvation in another. This new stream of thought had a marked effect on economics as on other doctrines. Was the break so sharp that present-day economics shows no family traits visible in the earlier ideas?

The reader may judge for himself as he scans the following pages. While obviously much has been added to economic thinking, and the aggregate of economic practices themselves has changed greatly, apparently much has remained in economics as in other branches of learning.

The attempt to apply scientific method has not altered the use of economics as an instrument of policy. On the contrary, the better the science, the better the policy.

Economic doctrines still vary with changing customs and needs, and still are used in controversies about policies and methods.

There is no such thing as economics in a vacuum. To have meaning at all, its doctrines must be considered in relation to a specific type of society, and in connection with aspects of behavior in that society other than merely making a living.

Economists still deplore degrading and stultifying labor, though instead of looking down on those who perform it they encourage technical progress, which substitutes machines and mechanical power for human slaves and beasts of burden.

Economics still must, and does, have to do with justice and moral codes.

Economists almost to a man still abhor avoidable waste and inefficiency. They believe that production should be for use, not for mere accumulation, no matter what the immediate incentive of the producer or what roundabout means is best adapted to the end.

Economists still do not believe that money in itself constitutes riches or that money is useful except as a servant of desirable production and exchange.

In a word, economics has remained the study of household management, be the household a family, a city, a farming community, a corporation, a nation, or a world.

To know something of the changes in economic doctrine orchestrating these basic themes will help in understanding economic problems of today. Capitalism *vs.* socialism, the "welfare state," inflation, taxation, prices, unemployment, and depression—all such problems may appear less obscure if one can bring to them such illumination as may be derived from leading economic ideas of the past and present.

2

The Nation as Merchant

The feudal system was based largely on the need of agricultural populations for protection against marauders—a protection provided by the lord for his retainers and serfs. Each manorial estate produced most of what it consumed. As Europe became more orderly, commerce by land and sea was made safer, and the need for protection diminished. Those with surplus crops could sell them, buying in exchange other farm products that they lacked or handicraft manufactures from the towns. Cities grew, merchants prospered, and a commercial system with its markets steadily dissolved the old custom of local production for use.

The growth of city populations, of making goods for sale, and of trade over longer distances was a major influence leading to the emergence of the nation-state as a center of power. In Spain, Portugal, France, England, and the territorial princedoms of Germany absolute monarchs reigned as a symbol and expression of state sovereignty. Unlike the feudal lords, the state employed a paid, professional army to safeguard its power, administered its affairs through paid officials, and had to collect taxes to meet its expenses. Like merchants, the state needed and valued money. The more money people had, the more the state could collect and spend. Therefore the state encouraged money-making occupations such as manufacture, trade, and banking.

The enterprising merchants and craftsmen needed the national state as much as the national state needed them. The crown built, maintained, and safeguarded highways and waterways and abolished private tolls on them. It encouraged groups of businessmen by subsidies and grants of monopoly. It protected the infant industries against foreign competition for their products on the one hand and facilitated the import or forbade the export of their raw materials on the other. Its authority broke through the feudal restrictions on trade and accumulation of wealth and leveled the barriers of the conservative guilds against entry of newcomers into their lines of business or competition with their products. In short, the national state provided nourishing food for the sturdy youth of capitalism, as it shouldered out the old feudal system with its privileges, class stratification, methods of production and distribution, and moral precepts. The new practices were not devised as a system: they grew like weeds in a newly cleared forest.

New World Too

Trade flourished not only within Europe: it extended over longer distances. During the Middle Ages themselves the Crusades had stimulated commerce with the lands east of the Mediterranean. But the overland caravan route to India was perilous and expensive. Now the Portuguese Vasco da Gama found a longer but cheaper way by sailing around Africa, and soon all the Atlantic nations were busy establishing profitable trading bases on the islands and coasts of Asia.

After the search for a shorter route to the Indies led to the discovery of America, another powerful stimulus was added to the feverish race for riches. Spain, which had financed Columbus, was of course first in the field and aroused the envy of the rest of Europe with its booty of gold and silver looted from the natives of Mexico and Peru. This type of wealth was certainly not useful according to the doctrines of Aristotle or St. Thomas Aquinas. But with it, favored Spaniards lived in luxury by buying extravagantly whatever they wanted, and the Spanish crown replenished its fortunes.

Thus more money was circulated throughout Europe at a time when it was needed to provide both a medium of exchange for growing commerce and more money-capital for investment in new enterprises, domestic and colonial. Indeed, more than enough money became available to finance the number of transactions, and so prices rose spectacularly. Europe experienced what we should now call inflation, with exaggerated profits to producers and hardships to those of fixed and small incomes. But the profits stimulated trade and facilitated new investment. The inflation dealt still another blow to the old "natural" economic order.

Exponents of the New Order—Machiavelli, Bodin, Serra

Though thinkers had not devised the growing new order of capitalism, they contributed greatly to its development and perhaps more important, elaborated doctrines which helped to clothe it with conviction and respectability, since in many respects it was opposed by the landed nobility and the churchmen.

Niccoló Machiavelli (1469-1527) was both a practicing statesman in Florence—a sort of brain-truster for the rulers of the city—and an intellectual leader of the Renaissance. In politics he rejected the moral codes of the learned clergy, which received their sanction from classical authority; instead he went back for inspiration to what he thought was classical practice. In *The Prince* he supported the idea of the supremacy of the state over all other sources of power, including the church. A benevolent despot, he argued, was above morality in his public acts provided he sought justifiable objectives. Necessary objectives for the state were, he declared, the extension of its own power and material prosperity.

A Frenchman, Jean Bodin (1530-1596), likewise wrote in defense of sovereignty while assisting the practical exaltation of the power of the sovereign. In the court of Henry III he sustained the monarch against religious factionalism. In his theory, which has exerted great influence in political doctrine until recently, the state has by right supreme power over the citizens, being above the law. Bodin's emphasis, like that of Machiavelli, was an exaggerated reflection of the actual need of an emergent national state for more power, in order to maintain order and facilitate the creation and accumulation of wealth.

Spain, with its direct access to gold and silver and its rough-and-ready method of acquiring them, did not take to manufacture or even much to trade and had no need to justify them. The Italian city-states, on the contrary, had few mineral resources. However, with their central position in the Mediterranean, they were experienced artisans and traders. A Calabrian, Antonio Serra (1580-1650), was one of the earliest to develop a theoretical argument for the benefits of commerce to a nation.

In *A Brief Treatise on the Causes which Can Make Gold and Silver Plentiful in Kingdoms Where There Are No Mines*, Serra maintained that manufacture was superior to agriculture for this purpose, since its products could more readily be sold abroad, thus bringing in money. Agriculture was de endent on the weather, whereas labor applied in trades was sure to bring gain." Crops were limited by the amount of and which could be devoted to them; not so with handicraft operations. Farm products were perishable and bulky and ould not profitably be stored for long or carried great distances, but woolen cloth, linens, silks, arms, pictures, sculpture, printing, drugs, and the like "may be exported with very facility to any distant country." Moreover, they yielded greater profit.

Bullionists and Thomas Mun

It was the natural assumption of the merchants, rulers, and thinkers of the time that the accumulation of money or the metals out of which it was chiefly made—gold and silver—must be the chief objective of national policy. Was not money the most convenient and tangible form of wealth? With it, one could buy anything else, for it was acceptable throughout the world. It was durable and could be stored indefinitely until needed. It embodied large value in small compass. Spain had led the way in becoming rich by accumulating gold and silver. Other nations sought, with what resources were available, to follow Spain's example.

The earlier advocates of this policy, since they concentrated narrowly on the desirability of gaining and holding gold and silver bullion, are called "bullionists." If this was to be the chief end of the state, and if the state was rightly all-powerful, the logical inference seemed to be that government should forbid the export of gold and silver and should try to maximize import of bullion by strict regulation of international payments and the individual transactions that brought them about. Thus a foreign trader who sold goods for precious metals to be brought into the country was to be encouraged, but one who merely bartered one commodity for another, or built up his capital abroad, was evading his duty. England forbade the export of specie and tried to control the dealings of traders.

The British East India Company was chartered by the Crown in 1601 to exploit trade with the East Indies—not only as an enterprise for the profit of those who put up the capital but as a chosen instrument of national policy. The company got into trouble when in 1613 one of its ships was wrecked, giving rise to the discovery that it was carrying a large amount of bullion out of the country. A spirited controversy arose out of this incident.

Among the defenders of the company was Thomas Mun 1571-1641), a man of education and wealth, son of a London merchant. He joined the East India Company in 1615 and acted as what would now be called a public-relations expert —certainly one of the most noted in history. His mature contribution to the theory of foreign trade, written about 1630 but not published until after his death, has become one of the classics of economics. It was entitled *England's Treasure by Foreign Trade.*

Mun argued that "the ordinary means therefore to increase our wealth and treasure is by *Foreign Trade,* wherein we must ever observe this rule: to sell more to strangers yearly than we consume of theirs in value." If, argued Mun, this rule is observed, the net result will be to bring money into the country, even if in the process of enlarging foreign trade some gold and silver have to be spent in foreign lands.

Suppose England has already sold abroad everything it can produce in a year for export; suppose, too, that the value of these goods is greater than the value of those it imports, and that England has collected the difference. How can England possibly earn any more money in that year? Easily, answered Mun, by buying with money, say, pepper in the East Indies for £100,000 and selling it in Italy or Turkey where it would bring £700,000. Though the difference would not all be clear profit, even the expenses paid by the merchant for shipping, wages, and so on would get into English hands.

Mun carried the argument a step further than this. Trading is profitable even if gold or silver be lacking with which to carry it on. "The Italians and some other Nations," he pointed out, "transfer bills of debt, and have Banks both public and private, wherein they do assign their credits from one to another daily for very great sums with ease and satisfaction by writings only, whilst in the mean time the Mass of Treasure which gave rise to these credits is employed in Foreign Trade as a Merchandize, and by the said means they have little other use of money in those countries more than for their ordinary expenses. It is not therefore the keeping of our money in the Kingdom, but the necessity and use of our wares in foreign Countries, and our want of their commodities that causeth the vent and consumption on all sides, which makes a quick and ample Trade."

Indeed, continued Mun, keeping too much money within a country is bad, since "plenty of money in a Kingdom doth make the native commodities dearer, which as it is to the profit of some private men in their revenues, so is it directly against the benefit of the Public in the quantity of the trade; for as plenty of money makes wares dearer, so dear ware decline their use and consumption. . . . It is a true lesson for all the land to observe; lest when we have gained some store of money by trade, we lose it again by not trading with our money."

In these passages Mun came close to the recognition that even in a commercial order money is wealth only when it is used as a medium of exchange, and becomes merely a public burden when it is hoarded. He recognized clearly that there can be too much money as well as too little for the public good. And he was on the verge of accepting the view that real wealth consists of the actual goods and services produced and consumed. In this he was in virtual agreement with defenders of the "natural" order, from Plato and Aristotle to the medieval churchmen. He differed from them essentially in seeing that profitable production and trade on a worldwide scale could yield more real wealth than local production for local use.

Balance of Payments and Foreign Exchange

Mun, having been actively engaged in foreign trade, learned from experience some of the facts about it that have been used by economic students of the subject from that day to this. His most important contribution was an elementary form of the account which is now called the international balance of payments (of any nation for any given year). Such a balance could not then be drawn up with much precision in the actual figures; even now some of the essential data are lacking. But he correctly set down the principles.

On one side of the account he placed the value of goods exported plus such items as are now called "invisible" exports —payments to Englishmen for shipping freight, insurance, and the profits received by merchants. On the other side he listed the amount paid abroad for imports. This should not include the profit made in selling them in England, the freight, insurance, import duties, and so on not paid to foreigners, for such items were paid by Englishmen to Englishmen. But there were "invisible" imports to be added to imports of goods, in the form of freight, insurance, duties, and the like paid to foreigners.

Finally, account must be made of other "invisibles"—such things as expenses of travelers, gifts, interest received or paid. Payments of this sort obviously counterbalanced part of the payments made for purchase of imports or sale of exports. When travelers came to England, for example, what they spent there helped Englishmen buy imports and so reduced the net amount Englishmen had to spend abroad.

When all relevant items were accounted for, the remainde
obtained by subtracting outgoing payments from incomin
ones must equal England's gain in treasure. "Let Princes op
press, Lawyers extort, Usurers bite, Prodigals waste, and lastl
let Merchants carry out what money they shall have occasio
to use in traffic. Yet all these actions can work no other effect
in the course of trade than is declared in this discourse. For s
much Treasure only will be brought in or carried out of
Commonwealth, as the Foreign Trade doth over or unde
balance in value. And this must come to pass by a Necessit
beyond all resistance."

Living economists could add little to this analysis of th
balance of payments except for one major factor, then les
important—the incurring of long-term international debts, o
the obverse of this factor, international investment. A natio
may continue to import more than it exports, without payin
the difference in gold, if the balance is accounted for b
foreign loans or investment coming in.

Thomas Mun also understood what some still regard as
mystery—the operation of foreign exchange. "The Merchant:
Exchange by Bills," he observed, "is a means and practic
whereby they that have money in one country may deliver th
same to receive it again in another Country at certain time
and rates agreed upon, whereby the lender and the borrowe
are accommodated without transporting of treasure from Sta
to State. . . . That which causeth an under or overvaluing o
moneys in Exchange, is the plenty or scarcity thereof in thos
places where Exchanges are made. For example, when there i
plenty of money to be delivered for *Amsterdam*, then shall ou
money be undervalued in Exchange, because they who take u
the money, seeing it so plentifully thrust upon them, do thereb
make advantage to themselves in taking the same at an unde
value." (By undervalue or overvalue of the pound or any othe
unit of currency, Mun was of course speaking in terms of i
value in weight of gold or silver.)

And what causes the oversupply of money in any exchang
market? Obviously it is the fact that the payments that mus
be made there are larger than those to be received—a disparit
which arises from the course of trade.

When an English pound could buy fewer Dutch guilder
Englishmen would have to use more pounds to make thei
purchases in Holland. This, contended the advocates of re
stricting the exchanges, meant loss to England. But, Mu
argued, the depreciation of the exchange in one market doe
not cause any export of money that would not otherwise occu

nd does not call for laws limiting the operation of the exchanges. England would be bound to lose money to any
market where there was a trade deficit, even if the two moneys
were not allowed to be exchanged at all. The Dutch, if they
had to pay to England, say, £400,000 in gold and could draw
from England, say, £500,000, would still net £100,000.

What matters, Mun argued, is maintaining an export surplus
with the world as a whole. If that were done, England would
be sure to take in more money than she paid out all together,
and deficits in particular markets would not be serious. It
would all come out in the wash.

How to Increase the Earnings from Foreign Trade

Clearly Mun was a good observer and a logical thinker. In
the realm which he knew best—foreign commerce—his observation and his thinking led him to advocate freedom from
state interference with the operations of a concern like the
East India Company. But he did not carry this conclusion to
the extent of advocating freedom for others. He never questioned his basic assumption that an export surplus should be
the chief goal of national policy. And he took for granted the
doctrine actually in practice and supported by his predecessors
that the state ought to play a large role in creating the conditions favorable to that goal, through the exercise of its sov-
reign power. Some of the measures he advocated for this
purpose were no doubt desirable on other grounds, but, as we
shall see later, some were not.

What could be exported was what could be spared by the
domestic population. Much could be done to restrict imports
and to expand the exportable surplus. If England would use
its waste land to produce hemp, flax, tobacco, and other crops
customarily imported, it would diminish the need to buy
abroad. Purchase of foreign-made luxuries could be decreased
by enforcing laws against "excesses" in consumption.

An effort to direct production into scarce goods most wanted
abroad would pay, as would underselling competing nations
in more common necessities. It would be desirable to restrict
to English ships the carrying of English commerce, and so
avoid the paying of freight to foreigners. Why buy fish from
the Dutch when they are caught in English, Scottish, and Irish
waters? Why not restrict purchase to fish of native fishermen?
Profit can be made by importing goods to be re-exported, and
to do this warehouses for foreign merchandise are desirable.
Trade with distant places like the Indies brings in more

money than trade with neighboring regions, hence should be encouraged, even if the trader makes no more profit at it. For example, pepper sells in London for two shillings (24 pence) a pound. It can be bought from Amsterdam for 20 pence—a profitable transaction. But if the merchant buys pepper imported from the Indies by an English company, only 3 pence a pound need be paid the Indians. Even if the price of the pepper delivered in London is still 20 pence, the costs of transportation, insurance, and the like are paid to English rather than Dutch persons and are a mere transfer within the kingdom.

Export duties (then customary) should be reduced. On goods manufactured out of foreign materials they should be abolished entirely. This would enable English manufacturers to compete abroad successfully with Italy and the Netherlands and would lead to more employment. Likewise there should be no duties on goods brought in to be re-exported. But taxes on domestic consumption of imported goods should be raised both to discourage payments to foreigners and to enlarge the king's revenues.

In general, industry ought to be encouraged rather than agriculture, since the more labor applied to any natural material, the greater its value and the more it would bring in foreign sales.

French Mercantilism—Colbert

In Thomas Mun's writings are found the principal arguments of those who regarded the nation as if it were an individual merchant. In France the great exponent of mercantilism was a statesman rather than a writer—Colbert. Consequently the policy is usually called *Colbertism* by the French. His aims were to serve the power and glory of the state rather than to increase private wealth, but the enhancement of national power and the seeking of gain were almost indistinguishable in the minds of those who favored the mercantile regime. Rivalry among the Spanish, the French, the English, and the Dutch, accompanied by frequent wars in the course of the struggle for power, was just as much a part of the new order as were its commercial policies.

Jean Baptiste Colbert (1619-1683), the son of a wool merchant, entered public service and at the age of thirty-two was chosen to manage the estate of Cardinal Mazarin. This he did with such success that Mazarin recommended him to the king, and he became minister of finance and virtual dictator under Louis XIV. He fostered manufacture and commerce by tariff on imports, bounties to French shipping, extension of French colonies, improved transportation at home. Since a plentiful and cheap labor supply was essential for his purposes, French workers were forbidden to leave the country and immigrants were attracted. Colbert granted monoplies to encourage new enterprises, especially in overseas trade, stimulated invention, established model industries. Within a decade he doubled the king's revenues, and made France the most powerful nation in Europe, with a mighty naval establishment. Science and learning were favored with academies, libraries, subsidies.

All this took public spending, and taxes soared. The king, in spite of his greatly increased revenues, spent even more freely than he received, and national bankruptcy neared. For obvious reasons, the people did not seem to appreciate power and glory bought at so high a cost, and Colbert, in poor health partly as a result of his skillful and unremitting labors, died intensely unpopular.

The Benefits from Mercantilism

Mercantilism was the name applied to this type of regime by subsequent critics of it, notably Adam Smith; the name was not adopted by its practitioners. They were not, indeed, a self-conscious school of economists at all, nor did they pretend to outline a complete and scientific system of thought. They were rather practical politicians, statesmen, merchants, who wrote in defense of favored policies or interests which in practice they were furthering.

In a sense the name is apt, since it implies that the best policy for a nation is the same as that for an individual merchant, who seeks to take in more by selling than he pays out by buying and so to build up the wealth of his firm. Like the merchant, the mercantilist nation regards others as competitors and tries to find ways of taking trade away from them by monopolistic or other devices. The merchant normally endeavors to beat down the price of what he buys, to pay as low wages as possible. Just so, the mercantilist state sought to grow wealthy by exploiting cheap colonial sources of supply and cheap labor at home.

Adam Smith's contention that state intervention did not promote the welfare of the inhabitants of a nation is extremely cogent, as we shall see in Chapter 3, but mercantilism nevertheless served useful functions in its day and comprised somewhat broader policies than the name invented by its opponents would imply.

In the first place, the doctrine of the bullionists that the aim of a state should be the accumulation of treasure in the form of gold and silver had some justification in the fact that trade over long distances and in impersonal markets was rapidly supplanting the household or manorial economy supplemented by barter. More money actually was needed as a medium of exchange when a monetary economy was replacing one in which markets played a minor role. To be sure, the need for money is not illimitable, and when its circulation grows more rapidly than the number of transactions in which it is used prices rise, as the mercantilists discovered. Yet few would dispute that a market system in which exchange is facilitated by money is much more favorable to the general advance of wealth and welfare than the more primitive system in which each locality has to produce most of its own necessities and can obtain the others only by exchange of commodities.

Money capital was also essential in financing the larger enterprises which came with the growth of manufacture and trade. Without it, joint-stock companies would have been nearly impossible.

Growth of the power of the nation-state provided security against the conflicts of warring principalities or nobles within it and formed an essential framework for wider internal markets and the arts of peace. Even the later opponents of mercantilism did not question this; few of them wanted to return to feudalism. In addition, modern private enterprise could scarcely have taken root without positive encouragement and aid from the state. Building of highways and canals needed and almost everywhere obtained, government support, if indeed these improvements were not wholly public enterprises. Monopolies and subsidies helped new industries get on their feet. Even protective tariffs, dubious though their benefits are in a highly industrialized world, may have been necessary at the beginning; no "underdeveloped" nation has built a large industrial base without them. State aids to science, invention, education, and other requirements for a technological culture have played an important role.

Colonial imperialism, now a discredited policy, was at first the only practicable way of opening up the new American continent. No private agency would have had the necessary resources, without government help, to carry on the exploration, begin the settlement, or furnish to colonists the means of living until they became established. Even in more populous regions like Asia, the extension of trade which was prompted by the mercantilist doctrine was a first step in breaching the barriers of distance and exclusiveness which for millennia had separated the East from the West.

As we look back on the broad range of events, it would be historically naïve indeed to assert that the birth of modern civilization could have taken place unattended by the nationalism in economic and political affairs which accompanied it.

The Sufferers from Mercantilism

The birth of modern civilization, however, like most great social upheavals, was far from painless. It became particularly uncomfortable for industrial labor, for those engaged in agriculture, and for exploited colonies.

Labor, which in feudal times had enjoyed a status of its own, however lowly, wherein rights were supposed to be combined with duties, became more and more a mere instrument of production for the profit of others. The wealth of a nation was not identified with the welfare of its population. The idea was to export as much as possible rather than consume it at home, and to get labor as cheaply as possible in order to promote foreign sales and the accumulation of treasure. Large populations were encouraged in order to augment the labor supply. If unemployment appeared, so much the better, since it tended to lower wages. Women and children clad in rags worked in English coal mines during the reign of the Tudors. Any tendency of workers to revolt was ruthlessly suppressed. Idleness was regarded not as a misfortune but as a crime; nobody could get relief under the Elizabethan poor laws without admitting to inferior status. Hours of work were virtually unlimited.

Agriculture, meaning in those days not the occupation of the actual workers on the land so much as that of the proprietors and managers, was explicitly relegated to a position secondary to that of industry, though for centuries it had been, and still was, the chief producer in Europe and the means of livelihood of a large majority of the population. Its function was regarded as that of furnishing food as cheaply as possible so that wages could be low, and industrial materials as cheaply as possible for manufacturers. In most countries protective duties on foodstuffs were removed while manufacture was protected. Landlords and peasants had to buy at high prices and sell at low ones. Though most mercantilist writers are silent on the point, Colbert and other statesmen tried in practice to prevent the export of grain, either by high export duties or absolute prohibition. In addition, landholders were subject to high and increasing taxes to support state activities.

Though industrial labor at the time had virtually no education and small political weight, the landed interests formed a powerful opposition to the new regime and had to be reckoned with.

In England many small tillers of the soil lost their land and their occupations because of the "enclosures"—seizures of land, legally or illegally, by large proprietors in order to devote it to sheep-raising rather than to grain, as before. Wool had become the most profitable crop because of the great expansion in markets for woolen cloth as trade and manufacture grew. The victims of this depopulation of the countryside thronged the towns, where they augmented the supply of cheap labor, became vagrants, beggars, and "criminals," or were herded off to colonize America.

Colonies were exploited for the benefit of the colonizing countries. In Spanish colonies Indians were enslaved and forced to work in mines or fields. The effort was to extract and export as much as possible of the gold, silver, and other natural wealth of the countries at as low a cost as possible. Negro slavery was first introduced in the West Indies and spread from there to South and North America. In the North American colonies under British rule, after the search for gold and silver brought no appreciable results, the effort was to use the colonies as sources of supply for materials absent or more expensive in England, and as markets for English manufactures, in order to increase the English "favorable balance of trade."

The mercantile policy was not in all respects unfavorable to the British colonists. They received subsidies for crops like indigo and a monopoly of the English market for tobacco. It is doubtful whether for many years they would have done much but export furs, fish, forest products, and later, crops, in exchange for British manufactures, even if there had been no restrictions on trade and production.

The Navigation Act of Cromwell in 1651 limited English imports to English-built, -owned, -commanded, and -manned vessels, but since the colonists were regarded as English, the shipping interests of New England shared in this monopoly. More burdensome were the later regulations, which channeled as many shipments as possible through London and limited American trade with other countries.

One continual complaint of the colonists may be directly traced to mercantilist policy: their assertion that money was being drained out of the country and they never had enough to carry on business, pay their debts, and finance expansion.

It was the expressed purpose of the mercantilists to draw money to England by a "favorable balance of trade." The colonial demand for money led to inflationary issues of paper bills and many controversies with English authority over this question.

More important than all specific sources of friction was the colonists' feeling of indignity because they were being deliberately used as instruments to serve the interests of the British realm, instead of being regarded as men with an independent right to seek their own welfare.

At first acquaintance it may seem that the mercantilist writers abandoned the moral outlook or the general views of former writers who touched on economics, to act instead as mere propagandists for kings and commercial interests. Yet they did not greatly differ in these respects from other schools of economic doctrine. They substituted a new moral code for an old one: the doctrine that state power was necessary in the public interest, as indeed in a measure it was. And though their writings did favor a special interest as opposed to others, they did not write with their tongues in their cheeks; they apparently believed what they said and strove to justify it with reasons of general appeal. Every body of economic doctrine arises from the needs or circumstances of its time; each may be used—or misused—by some group of special advocates. Each, too, contains some lasting truth.

3

The Classical Economists

Ideas were flowering in great profusion during the eighteenth century. Many influences combined to stimulate them: the growth of cities, which brought more people together and favored varied interchange of opinion; the increase of wealth; easier travel in a larger world, which broadened horizons; the growth of science; and the search for new systems of thought to replace the old.

The Enlightenment in France and England became a focus of this intellectual ferment. Though no brief summary can do justice to its welter of ideas, a few prevailing drifts of opinion proved to be of special importance to economic doctrine.

Beginnings of Modern Science

During preceding centuries people had depended on the ancients like Aristotle and on the church fathers for knowledge of the external universe. It was necessary only to consult the authorities and to deduce from their writings the explanation of any problem. Logic-chopping deduction had served instead of careful observation, new insights, experiment.

But various sturdy-minded men had now gained new and more accurate knowledge by humbly and objectively studying nature itself. Recognition that the earth was not the center of the universe but resolved about the sun, Harvey's discovery of the circulation of the blood, Newton's statement of the laws of gravitation and motion, were being followed by dozens of lesser but significant and startling observations.

If the old authorities were wrong about the physical universe, were they not wrong also about religion and the codes of human behavior? Everything became subject to question. Science was then called philosophy, and there was no distinction between the realms each was supposed to examine. The philosophical writers began to examine human institutions anew, just as they had examined non-human things. In doing so they tacitly assumed that man was a part of nature instead of a separate kind of being divinely ordained and cared for. Explanation of causes and effects in human behavior, desirable or undesirable, was to be sought in laws of nature rather than in the will of God, as expounded by sacred literature or the doctrines of a living church. Reason rather than authority was to guide.

Change and Progress

Medieval doctrine, if not always medieval practice, had been dominated by other-worldliness. The mundane sphere, including human life itself, was just a training ground for life after death, with its punishments and rewards. Suffering was to be endured with the knowledge that it was but a prelude to glory in a future life. There was little intellectual incentive to reform social customs or increase temporal well-being, except for whatever spiritual benefit might be gained.

Now the emphasis shifted to the improvement of life on earth for its own sake. Material gains were obvious on every hand; rapid change was in fact taking place. Science and invention opened up limitless possibilities of better or easier ways of doing things. A spirit of adventure was abroad. Could not philosophy do as well with human institutions as with material things? All that was necessary was the application of reason to man's ways of living together. Plans for utopia were designed by many. (The word utopia itself is derived from the title of a book by an Englishman, Sir Thomas More.)

Deification of Nature

Those who abandoned belief in God altogether had to find a substitute: they found it in Nature. Those who still adhered to religion—as most did in words even if not in fact—believed that God, instead of expressing his will directly, was bringing it to pass through the realm of nature and natural law. Nature thus became not just something that existed, but also something to be obeyed. To act contrary to nature was to be unpious or immoral. We still use the words "un-natural behavior" in this sense.

The effect of this view on doctrines dealing with human behavior was curious indeed. To a modern scientist a law of nature is not a command which the universe or atomic par ticles can disregard only at their peril. Rather it is an abstrac tion or generalization of the way such things *do* behave. If i a substantial number of cases or in critical instances exception are found, the scientist does not exhort the offenders to men their ways; he abandons or modifies the law, since it no longe offers a valid explanation of what he has observed.

But the philosophers of the Enlightenment tended to ac count for human miseries on the ground that men were dis obeying natural laws. All that was necessary for improvemen of social institutions was to discover the laws of nature tha should be applied, and then to obey them. Nature was re garded not only as orderly but also as benevolent. It was, i other words, still a kind and just Father, even if a somewha impersonal one.

It was but a step from this assumption to the idea that th "ills that flesh is heir to" had arisen from an accumulation o unnatural practices which had corrupted a primitive an therefore natural community. Writers harked back to som imaginary early state of society where everybody was happ and good because nothing artificial had been introduced. I this disguised form they revived, without being conscious tha they were doing so, the legend of the Garden of Eden as th prototype of a new earthly paradise which they were inten on creating.

Natural Rights and Individualism

Some declared that the "natural" original community wa one in which all the bounties of nature were held in commor and shared according to need, and hence that the new societ should be a communistic one. Others emphasized the indi vidual.

Hugo Grotius, the Dutchman, in 1625 published *De jur belli et pacis* (*The Law of War and Peace*), in which he de duced from the "originally social nature of man" the existenc of "inalienable and indestructible rights of the individual." Thomas Hobbes, the English philosopher, assumed that ir the natural state all individuals were free and dependent or themselves, but his view of man's nature as expressed in *Le viathan* (1651) was less optimistic than that of Grotius. Men he thought, naturally fought each other in the struggle fo existence rather than being cooperative. To escape the result ing turmoil they consented to the establishment of the abso lute state, delegating their natural rights to it by a tacit "socia contract."

John Locke (1632-1704), the chief English philosopher of the Enlightenment, emphasized natural rights of life and liberty, which underlay the movement for representative government. He added to them a natural right of property, since, he contended, no natural product had value except by virture of the labor expended on it, and every man had a right to the fruits of his own labor.

Jean Jacques Rousseau, the eloquent but inconsistent Frenchman, was a mighty force in the surge of ideas which helped to foment the French Revolution. His major works, *Le Contrat Social* and *Émile*, appeared in 1762. They, like the writings of Locke, were familiar to Thomas Jefferson and the other authors of the American Declaration of Independence, which asserts the "inalienable rights" bestowed upon man by "nature and nature's God," among which are "life, liberty, and the pursuit of happiness." Rousseau declared that in the state of nature men were good, free, and equal, and had become otherwise only through miseducation.

Anthropology reveals no such state of nature as the writers of the eighteenth century imagined. Primitive societies exhibit almost infinite variety. Ideas of what is right and natural vary widely according to the social settings in which they are found. Yet, partly because our own culture is derived from that of Western Europe, where ideas of individual natural rights exerted great influence, they arouse strong emotional responses within us. Emphasis on the dignity and autonomy of the individual, which is part of our tradition, seems natural and necessary now, as it did to those who formulated the ideas. The longing for freedom which the thinkers found in existing men was there; they were mistaken simply in transposing it to an imagined primitive state.

More doubtful, however, is the inference based on such doctrines by some political and social philosophers that all human institutions are or should be the result of purely individual self-assertion or self-protection. Society is not merely an aggregation of self-determined human atoms.

The Physiocrats' Attack on Mercantilism

The first modern school of thinkers to call themselves economists, to regard their theory as objectively scientific, and to develop a complete and self-contained view of the economic order as a whole, arose in France not long before the Revolution. By its later adherents this school was named Physiocracy, "the rule of nature." Like other thinking of the time, it was deeply influenced by the concept of natural law as both basic and benevolent.

The Physiocrats are notable as well for two other reasons. One was their invention of the term and policy laissez faire, which has lingered as a subject of economic discussion ever since and characterizes the doctrine of the classical economists who followed. The second was their quaint but groundbreaking analysis of the circulation of wealth, which purported to show how what we should now call the national income originated and was distributed.

The founder of the school, which made a frontal attack on mercantilist policies, was appropriately of rural origin and spent his early years on a farm. François Quesnay (1694–1774) studied surgery and qualified as a Doctor of Medicine at the age of twenty-five. So great was his skill and reputation that he became court physician to Madame de Pompadour, the king's mistress, and later to Louis XV himself. With his intellectual eminence and knowledge of science, he attracted prominent men at the court to conferences in his apartments, where he promulgated his ideas about economic affairs and hammered out a consistent theory.

Among those who participated were Mirabeau the Elder, Anne Robert Jacques Turgot, who later became for a while finance minister under Louis XVI, and Pierre Samuel Du Pont de Nemours, who invented the name Physiocracy. Du Pont as a nobleman was condemned to the guillotine during the Revolution, but escaped death through the timely end of the Reign of Terror. He emigrated to Delaware, and later was commissioned by President Jefferson to draw up a system of education for the United States.

Physiocracy returned to the traditional doctrine that all wealth was derived from the land. Without food and fibers, wood, minerals, and stone, man could not exist, to say nothing of accumulating possessions. The husbandman was therefore the only true producer. Industrialists, traders, craftsmen, were "sterile." The land-owners, who directed extraction of wealth from the land and apportioned its use, were the proprietary or distributive class. Their responsibilities in a natural order were of prime importance, since they were guardians of the only economic surplus. This doctrine was obviously a sympathetic one to the agriculturalists who suffered from mercantile emphasis on trade and industry.

The curious belief that city workers were not productive was supported by the following line of reasoning. Only the earth could yield more than was put into it. Seeds, becoming crops, returned manyfold. Animals, breeding and reproducing themselves, rapidly increased in numbers. In a natural society, which was conceived as a primitive one based on agriculture, the occupant of land could, if he wished, tan his own hides, make his own shoes, spin and weave his own clothing. He had no need of artisans or traders. But in the course of time it suited his convenience to engage others to do such things for him, since specialists in each craft could do better jobs than one man who tried to learn all the arts of fabrication.

The surplus above his own needs harvested by the farmer enabled him to feed, clothe, and house the craftsmen whom he commissioned, passing out to them what was necessary to satisfy their wants. But the craftsmen were not in a position to accumulate wealth unless some "artificial" agency like the state intervened on their behalf. As for propertyless wage-earners, they could receive only what was necessary to keep them alive, since there was always competition for jobs. This view might have been inhumane, but it accorded with the generally observed conditions of the time.

A society in which most of the land was owned by those who did no manual labor, while others planted and reaped their crops, could not be regarded as natural in the sense of being primal, but it was, contended the Physiocrats, a natural and inevitable development. The less intelligent, lazy, and prodigal farmers in the course of time lost their land, which was bought by the good farmers who had built up a surplus. Eventually estates became too big for a single man and his family to work, and so he got the landless to do it for him by supporting them.

One cannot accuse the Physiocrats of emphasizing only that part of their doctrine which served a special interest. Since the only surplus arose from the land, only the land, they argued, should be taxed. All other taxes, like excises and import duties, were interferences with the natural order; they should be abolished. The proposal of a "single tax" on land has echoed through economic thinking ever since; late in the nineteenth century it was supported for somewhat different reasons by the American Henry George in his *Progress and Poverty*.

Laissez Faire

The Physiocrats contended that everything which represented a departure from the "natural order" was the cause of dissatisfaction and confusion and should be done away with.

Why should national policy strive to produce an export surplus, often called a "favorable balance of trade"? This did not result in the accumulation of any real wealth: wealth could arise only by skillful use of native natural resources. In reality, an export surplus meant merely that the nation was shipping out more goods than it was receiving in return. Such a process would impoverish any farm or household. If an export surplus was obtained by taking people off the land to fabricate exportable goods, the nation was merely draining its own lifeblood.

To favor industry by granting a monopoly was to give non-producers the means of extorting, by high prices, the wealth naturally belonging to producers. A state subsidy was even more barefaced robbery. Corruption, favoritism, and state bankruptcy thrived on such practices.

In a natural order, Quesnay argued, the prices of industrial goods would be higher than those of the materials out of which they were made only by as much as the cost of what the labor devoted to them had to consume. Prices, therefore, were, or should be, based on labor cost. This view was a forerunner of the classical labor theory of value.

Individuals ought to be free to act in their own self-interest. They ought to be allowed to choose their own occupations, to move about, to gain wealth, and to do what they liked with their property. The state should neither hinder nor help them. This was in a sense a moral law—the natural law of individual rights. It was also suited to the general advantage, since it would inevitably (being a law of nature) work out well.

All these doctrines were summed up in the slogan, "*Laissez faire et laissez passer, le monde va de lui-même.*" This may be somewhat liberally rendered as, "Don't interfere, the world will take care of itself." A comforting doctrine indeed, but one that could be accepted wholeheartedly only by those who thought they were in a position to get along nicely if others would only leave them alone.

Strangely enough the Physiocrats combined these opinions with a belief in absolute monarchy. Yet, on reflection, it does not seem so strange. Since natural law was not being observed in practice, there had to be someone to enforce it. The king himself was supposed to be morally bound by natural law to see that it was carried out. The Physiocratic position is very like that of the present-day economist who favors competitive free enterprise and the absence of state intervention in one breath, while with the next he calls for strict anti-trust legislation and a powerful government to enforce it.

The Economic Table

Quesnay's *tableau économique*, though now merely a curiosity, was thought by Mirabeau an invention as important as hat of printing. The idea behind it was indeed important, ince it involved a concept similar to that now known as the national income, a major tool of economic thinking. It purported to illustrate the self-contained circulation of income within any national population.

The table, representing expenditures and receipts for a year, consisted of three columns. In the central column, corresponding to the strategic position in the order, stood the landlord. The column at his left represented "productive" expenses— payments to the husbandman—and the column at his right "barren" expenses—payments to the manufacturers, etc.

The landlord started with a sum, let us say, of 2000 livres, or French pounds. He advanced 1000 to those who did his farming, a payment necessary to sustain them in producing the next crop. The other 1000 the landlord advanced during the year for products of manufacture. Both farmers and manufacturers spent the money received, buying what they needed, the manufacturer's 1000 livres going to the farmer, the farmer's going to the manufacturer. As a result of these reciprocal payments, made through the medium of the landlord, it turned out that at the end of the year he had received back his original 1000, while the farmers and manufacturers had each received and spent 2000. Everybody involved had been supported by this process, and in addition the landlord was replenished with the store of wealth necessary to start the next year's round of production and exchange. The land had produced the necessities of life, plus a surplus.

If the Physiocrats had carried this calculation one step further, they would have added together the annual receipts of all concerned, arriving at a total product or expenditure of 6000 livres. This income would have been "generated," as one school of contemporary economists would say, by the original expenditure of 2000 on the part of the landlord. The chief difference in conception would be that the 2000 of income-generating expenditure would now be called "investment." The school of economists in question (followers of John Maynard Keynes) would think of the investment as having been made by businessmen or government rather than just by landlords.

Physiocracy in Practice

The most prominent of the Physiocrats, Turgot, had an opportunity to try out their theory. Descended from an old Norman family, he was the son of a high government official and was educated in Paris. He expected to enter the church and won the degree of bachelor of theology. Instead, he went into public service and interested himself in economics. In 1761 he was put in charge of the distressed province of Limoges. There he introduced the culture of clover and potatoes and founded the famous porcelain industry of the region to make use of the local raw materials. He abolished heavy taxes on trade and organized relief for the destitute suffering from a year of famine.

This record was so good that the national administration, rapidly approaching bankruptcy and beset by troubles on every hand, called him to Paris as finance minister. There he tried to suppress the power of the trade guilds and so reduce the prices of necessities. Rigid economy in governmental spending was his program; he cut out subsidies and sinecures. At the same time he removed duties on grain and other hampering restrictions. The credit of the government was restored, but his reforms aroused the bitter enmity of the powerful interests he opposed, and Louis XVI, much against his will, was persuaded to dismiss him in 1776. "Only Turgot and I," the unfortunate king is reported to have said, "really care for the people." Turgot died before the Revolution broke out; his difficulties illustrate the obstacles usually confronting those who try to force people to obey economic "natural law" contrary to their special interests. Laissez faire, however attractive in theory, underestimates the ingenuity of those who can gain by establishing a system of laws of their own. It probably also overestimates the beneficence of the "nature" invoked.

Adam Smith, Classical Patriarch

The real founder of the classical school of economics, whose writings on the subject have probably had a wider influence over a longer period than those of any other, was the Scotsman Adam Smith (1723–1790).

An unusual combination of circumstances provided the setting for Adam Smith's great success. One was the new liberal doctrine of the Enlightenment, which not only inspired the Physiocrats in France but was sweeping all before it in advanced intellectual circles in Britain, where one of its sturdiest recruits was the Scottish philosopher David Hume, for years a close friend of Smith. Smith himself traveled in France and attended some of the Physiocratic conferences in Quesnay's apartments. The ideas springing from John Locke, Rousseau, and others had great influence also among the leaders of the American Revolution.

Another favorable circumstance was the fact that Smith lived at the beginning of the Industrial Revolution, when machinery, at first driven by waterpower and soon to be activated by steam, was multiplying production with great rapidity. The mercantile revolution had long since established itself, but the chances of gain in manufacture and trade were now being redoubled for those who knew how to make use of the many inventions which had begun to emerge from the new science and technology.

It was in Britain that the Industrial Revolution first took hold and achieved its dominance. The new regime of private enterprise in manufacture by power-driven machinery needed a system of ideas to justify it, not now so much against the landed aristocracy, as had the merchants, but against hampering restrictions and monopolies of mercantilism itself. Just as the Physiocrats played the role of spokesmen for agriculture in revolt against mercantilist privileges, so Adam Smith and his followers played the role of advocates for the rising industrialists and the exploited colonies.

Smith brought to this opportunity just the talents needed to make the most of it. He was a keen observer, capable of learning from the world of action even if what he learned did not comply with authoritative doctrine. He had a salty shrewdness and a taste for the illuminating example which made his writing readable by all, and specially appreciated by practical men. At the same time he exhibited a philosophical bent, strongly colored by ethical motives, characteristic of many Scots, which enabled him to cast his ideas in the form of a logical system which could be supported on the highest of grounds. Like most great economists, Smith was not a technically trained specialist but a thinker who came to the subject from a broader background and with fresh insights. The greatest fault of his doctrine, it was later said, was that he thought that inside everybody there was a little Scotsman.

How Smith Became an Economist

Adam Smith was the son of an official, born in Kirkcaldy, near Edinburgh. At fourteen (not an early age for college students at the time) he entered Glasgow University, then about the size of a present-day small liberal-arts college in the United States. There he studied, besides the classics and mathematics, philosophy under Francis Hutcheson, an apostle of liberty under a benevolent God. Hutcheson is said to have originated the ethical test of "the greatest good of the greatest number."

After three years at Glasgow, Smith went to Oxford on a

scholarship for a six years' stay, during which apparently he did not find life pleasant. He had intended to take orders in the Scottish Episcopal Church but gave it up. Returning to Edinburgh, he was soon offered a lectureship in English literature. In his lectures he sponsored some quaint opinions, including one current at the time that Shakespeare was not a good playwright. But the lectures brought him acclaim, and in 1751 he was called back to Glasgow to teach logic. The next year he received the post of professor of moral philosophy. It was from this background that he began to deal with political economy. A book on ethics, *Theory of Moral Sentiments* (1759), established his reputation. While at Glasgow, Smith became a companion of James Watt, credited with invention of the steam engine. In 1759 he supped with Benjamin Franklin and other learned men.

Smith's professorship turned out to be a steppingstone to an appointment which later helped him in conceiving and writing his great book, *An Inquiry into the Nature and Causes of the Wealth of Nations,* usually called simply *Wealth of Nations.* He became the traveling tutor of the young Duke of Buccleuch, at a salary of £300 a year—worth much more then than now—plus traveling expenses.

The tutor and his charge went to France, where they spent several years and became acquainted with leading French intellectuals and statesmen, including Quesnay, Du Pont, and Turgot. It is likely also that Smith met Thomas Jefferson. He began to work on the *Wealth of Nations* in Toulouse in 1764. That this classic of economic liberalism happened to appear in 1776, the year of the American Declaration of Independence, is not a mere historical coincidence. Similar forces led to both famous documents.

By the terms of his contract Smith received his salary as a pension for the rest of his life after returning from France —an endowment that enabled him to finish his work and enjoy the fame it brought him.

Edmund Burke said that "in its ultimate results" the *Wealth of Nations* was "probably the most important book that has ever been written." Burke's comment reminds us that in spite of Smith's penetrating observations of actual economic institutions, he was in reality a propagandist for a future liberal utopia—one which, he hoped, was just coming into sight when he wrote. Though some of his maxims were subsequently observed in practice, they were never completely applied anywhere, and the world today is further than ever from adopting his doctrine as a whole.

Major Aspects of Smith's Economics

The *Wealth of Nations* set out to explain how the wealth of a nation is increased and how it is distributed—the basic themes of modern economics.

The only source of wealth, he maintained, as we do today, is production resulting from labor and resources. In this he took basic issue with the mercantilist doctrine that a nation's wealth is derived from an excess of exports. Wealth will be increased according to the skill and efficiency with which labor is applied, and according to the percentage of the population which is engaged in applying it. The economic welfare of the average individual depends upon the relation of total production to the size of the population, or, as we should now put it, real income per capita.

These ideas, like many of Smith's, are so much a part of our tradition that they seem commonplace. But they were not commonplace when he set them down.

The basic means by which production is increased, Smith pointed out, are division of labor and the introduction of machinery. His famous passage on how the specialization of operations in the making of pins enlarges output per worker is a model of exposition. But the chance to apply a high degree of specialization depends on wide extension of the market, since large numbers of a given product cannot be sold in a small or local community, though this large output may be made by relatively few. Smith correctly observed that improvements in transportation, by widening markets, made possible the growth of commerce and industry.

Unlike Plato, he did not think that diversity of occupation arose from differing talents, but rather that differing talents resulted from diversity of occupation. The real cause of the tendency to specialize in production, he held, was man's innate propensity to barter and trade, which he alone among the animals exhibits. "Nobody ever saw a dog make a fair and deliberate exchange of one bone for another with another dog."

The fundamental role which Smith assigned to markets led him to speculate on how markets operate. The real or natural value of anything, he thought, was measured by the labor which would have to be devoted to making it. Nobody would take the trouble to make anything unless he thought it worth while. If he could buy something he wanted at less cost than the labor of making it himself, then he would buy it, giving in exchange something that the other participant in the swap could buy (in terms of labor) at less cost than *he* could make it. This gave rise to the mutual gain from specialization and trade. The natural value of anything depended not merely on the amount of time required to make it, but also on the intensity of the labor, the training or education which underlay the skill of the worker, and like factors.

But in an economy which uses money, the market price of commodities (or nominal value) was not always equal to the real value. The effective demand for any article—that is, not just the desire for it, but the desire backed up by willingness and ability to pay for it—sometimes exceeded the supply. This would increase the price and bring added gain to the producer. But the profitableness of the transaction would soon lead others to compete for the trade; it would draw labor and capital from other occupations. This would in time reduce the price, perhaps below the real value. Nevertheless, there would be a tendency for the price of any article to vibrate around, or to approximate, the real value. When demand and supply at any price were in equilibrium, then that price would represent the natural one. Thus, everyone would gain from a free market, since each would obtain what he wanted most at the lowest possible price; each would busy himself at what he could best do; productive resources would be apportioned according to consumers' wants.

When a man directs industry so that its product will be of the greatest possible value, "he intends only his own gain, and he is in this, as in many other cases, led by an invisible hand to promote an end which is no part of his intention. . . . By pursuing his own interest he frequently promotes that of society more effectually than when he really intends to promote it."

In another connection Smith wrote, "All systems either of preference or restraint being taken away, the obvious and simple system of natural liberty establishes itself of its own accord."

Smith's emphasis on individual self-seeking as the mainspring of social benefit, the beneficent "invisible hand," and the desirability of "the obvious and simple system of natural liberty" have been slogans of the defenders of private enterprise from 1776 to the present.

Distribution of the Product

The main ways in which money is paid for production and distribution are, according to Smith, wages, profit, and rent. His observations of what actually happened in these realms, while favorable to capitalism, candidly revealed some serious flaws in the "obvious and simple system of natural liberty."

In order to command the benefits of machinery, which increased wealth, consumers had to make it worth while for someone to save and invest his money in "stock," or capital, as we should call it. Thus the wage-worker could not receive the full natural value of the product, since some of the price had to be apportioned for profits. Profits were a necessary cost of production.

The more capital there was in the country, the higher its wages could be, since by accumulation of capital the nation was becoming more productive. Facts seemed to support this contention. Smith found reasons to believe that real wages had increased in England as industry had grown, and he correctly pointed out that there was a correlation between low levels of living in India or China and their lack of industrial development.

But wages were not necessarily as high as they might be, because of competition of workers for jobs. The lower limit of wages, Smith thought, was a minimum of subsistence, since below this level the working population would die off. Scarcity of labor combined with rapid increase of national wealth, which he correctly noticed as a rule in the North American colonies, led to high wages. Wages, like other prices, were subject to the law of supply and demand.

The poor, he commented, had an extraordinarily high birth rate, as well as a high rate of infant mortality—a truth confirmed by many subsequent statistical studies. The high birth rate added to the labor supply and tended to keep wages close to the subsistence level in static economies, or even below it in retrogressive ones. The necessary condition for rising wages was an *advancing* economy—that is, one in which ability to produce outstripped the growth of population. On the other hand, the prosperity of the rich led to a low birth rate and so tended to concentrate wealth among those who had it. "Luxury in the fair sex, while it inflames perhaps the passion for enjoyment, seems always to weaken, and frequently to destroy altogether, the powers of generation," wrote this perceptive bachelor.

Smith noted, too, that there is an inequality of bargaining power between wage-earners and employers. "Masters are always and everywhere in a sort of tacit, but constant and uniform combination, not to raise the wages of labor above the actual rate." These combinations were seldom heard of by the public, but attempts of labor to combine and resist reductions or gain increases were cause for outraged comment, and the violence which sometimes accompanies such movements was rigorously suppressed.

As for profits, their rate tended to decrease as any nation accumulated capital, since more capital meant more competition in industry. Extremes of wealth and poverty were more marked in backward societies.

Smith did not fail to note the necessity for competition as a condition of his system of natural liberty. He was an inveterate enemy of monopoly, except for a few functions where it would be wasteful, as in the building of canals. Monopoly, he wrote, "is an enemy to good management, which can never be universally established but in consequence of that free and universal competition which forces everybody to have recourse to it for the sake of self-defense."

Rent, Smith argued, was in essence a monopoly price. The quantity of good or desirable land is limited, and those who own it can extract something from the consumer which is a payment neither for labor nor for necessary capital. High rent is merely a result of great national wealth or high wages. In his analysis of rent, Smith foreshadowed the influential doctrine of the "unearned increment." In it, he also reflected the feeling of the industrial producer against the survivals of the feudal system and the landed proprietors.

Adam Smith also had much to say about money and interest, but this is a more technical subject in which his contribution was not so noteworthy. He agreed with almost all other authorities that large additional quantities of money increased the general price level; he regarded money as useful principally as a medium of exchange but saw nothing wrong in charging interest for loans. Interest would be low, he thought, when savings were abundant, high when the demand for money capital exceeded its accumulation.

Differences from Previous Doctrines

One of the most trenchant parts of Smith's book is his attack on the mercantilist doctrines, following inevitably from his views. If competition, freedom of trade, and specialization were desirable within a nation, they must be equally desirable among nations. No nation can gain by making something which it could buy more cheaply elsewhere. Yet to effect that extravagance is just the purpose of tariffs and subsidies. Without them, each nation would naturally specialize in the types of production for which it would be better qualified. In any case, its own production is the only source of its wealth, and the more of value it can produce with the labor available, the richer it will be. Monopolies affecting international trade are just as undesirable as those operating at home.

As a practical Britisher, Smith admitted exceptions. For purposes of military security it was desirable to promote British shipping. And tariffs should be laid on imported goods when the domestic production of the same goods was subjected to higher costs by a special tax, so that competition might be equalized. Tariffs could also be used for bargaining purposes. Finally, reduction of tariffs might justifiably be made by "slow gradations" instead of all at once, if undue hardships would be caused by sudden change.

The central charge against mercantilism Smith summed up by saying that it sacrificed the interest of the consumer to that of the producer, whereas "consumption is the sole end and purpose of all production."

With Physiocracy, Smith had more in common, but he criticized it as an overreaction to Colbertism. "If the rod be bent too much one way, says the proverb, in order to make it straight you must bend it as much the other." This the Physiocrats did in representing agriculture as the sole source of wealth. Smith thought agriculture of prime importance, but he regarded industry and trade as productive too. (Strangely, he did believe that personal servants, actors and artists, clergymen, and other purveyors of non-material services were unproductive!)

Quesnay, Smith thought, was too rigorous in his views of "natural liberty" by advocating promotion of agriculture at the expense of industry. "Some speculative physicians seem to have imagined that the health of the human body could be preserved only by a certain precise regimen of diet and exercise, of which every, the smallest violation necessarily occasioned some degree of disease." Yet the human body was capable of health "under a vast variety of different regimens." Quesnay, who was himself a physician, and a very speculative physician, made the mistake of prescribing too precise a regimen for the body politic.

What was Adam Smith's conception of the role of government, or, as he called it, "the Sovereign or Commonwealth"? A very simple one.

The government must provide for national defense, but in doing so must have regard for the economy which supports the armed forces as well as for the military itself. Industry is the essential basis for military strength, and becomes more so as arms become more highly developed and expensive.

The government should dispense justice, but there should be an independent judiciary.

The government should bear the expense of public work advantageous to all, though some of them may be made self liquidating through the payment of tolls.

The government should protect foreign commerce in general but should not perpetuate monopolies or delegate armed defense to particular interests, as Britain did in the case of the East India Company.

The government should subsidize elementary schools for the common people.

And the government should maintain the dignity of the sovereign by supporting the style of living expected of him.

Taxes to pay these expenses ought to be levied on the people "in proportion to the revenue which they respectively enjoy under the protection of the state"—for example, by an income tax. "The time of payment, the manner of payment, the quantity to be paid, ought all to be clear and plain to the contributor, and to every other person." Anything else leads to injustice or extortion. "Every tax ought to be levied at the time, or in the manner, in which it is most likely to be convenient for the contributor to pay it." Taxes should not cause large sacrifices to the taxpayer over and above the revenue which the taxes bring in. Taxes which discourage trade and industry are of this nature.

These prescriptions for government are noteworthy more for their omissions than for their inclusions. The government of the day, especially in England and France, enforced extensive regulations of trade both domestic and foreign, granted favors and monopolies liberally, imposed multifarious taxes both direct and indirect. Sinecures, favoritism, and corruption were rampant. Heavy military expenses were incurred in wars occasioned in part by exploitative and imperialistic colonialism. The American War for Independence, brewing when Adam Smith wrote, was in large measure a protest against just such practices.

The Classical Pessimist—Malthus

The influence of Adam Smith, great though it was, was almost equaled by that of Thomas Robert Malthus, who belonged to the next generation. He was born in 1766, ten years before Smith published the *Wealth of Nations*, and died in 1834. His early life was thus lived in a period of turbulence and revolution.

Two major circumstances influenced his thinking: one, the

many millennialist doctrines; the other, his observations of human misery. The Industrial Revolution was not at the time so beneficial to workers as Adam Smith thought it would be. There were frequent depressions and crises; bitter and unrelieved unemployment was caused not only by the cyclical instability of the new economic regime, but by the fact that machine industry was rapidly displacing handicraftsmen. The market for goods did not expand so rapidly as the labor supply. Many believed that England was overpopulated.

Either human beings were displaying an insuperable reluctance to obey "natural laws," or nature was not so benevolent as Smith and his predecessors supposed. Malthus was intrigued by the proposals of those who believed in the perfectibility of human nature, but he was equally impressed by obstacles in their way. As an ordained minister, he felt impelled to find some rational explanation for the oppressive social problems of mankind, so that a valid remedy might be found.

The Malthusian Doctrine

Malthus's *Essay on the Principle of Population*, published anonymously in 1798, caused a heated controversy. He then dug deeper into the subject, accumulated voluminous evidence, and in 1803 brought out a new edition much longer than the original, this time under his own name. Four more editions were published before his death. Malthus, stressing the struggle for existence, helped to form the ideas of Charles Darwin and Alfred Russel Wallace, who first formulated the theory of evolution.

The Malthusian doctrine is simple in essence. Unchecked breeding of man causes population to grow by geometrical progression, whereas the food supply cannot grow so rapidly, or more than in an arithmetical ratio. "I think," wrote Malthus, "I may fairly make two postulates. First, that food is necessary to the existence of man. Secondly, that the passion between the sexes is necessary, and will remain nearly in its present state." The continual pressure of population on subsistence had historically been relieved only by war, pestilence, and famine.

Real wages could not rise much above the level of subsistence because an increase in well-being would lead to a larger supply of workers. And when wages fell below this level, the surplus would be eliminated by death. This conception has become known as the "iron law of wages."

Malthus, like Smith, noticed the high wages in America

but attributed them to the high ratio of land to the number of inhabitants. Population was growing there with extreme rapidity, not so much because of immigration as because of a high birth rate. "It may be expected that in the progress of the population of America the laborers will in time be much less liberally rewarded."

Though concerned mainly with supporting his main thesis, Malthus contributed an idea which has been assimilated into general economic theory and has become one of the major tools of the classical type of analysis—the "law" of diminishing returns. He developed it in relation to land. A given piece of land will yield more with the application of fertilizer and the use of more labor. But there comes a point beyond which it does not pay to add to the effort to improve land. Additional increments of expense do not correspondingly increase the yield. Carried far enough, they will reduce it. The same reasoning was later applied to industrial establishments, even to whole industries.

Remedies Proposed by Malthus

Being a clergyman, Malthus was sensitive to the charge that if his theory was correct it was impossible to conceive of a good and benevolent God. Man, he replied, could apply the remedy by prudence and self-restraint, just as he could avoid illness due to gluttony or drunkenness. He advocated late marriage, which he argued would be good for human character and for the institution. Everyone should resolve to have no more children than he could support.

This moral code should be reinforced by society, through the simple expedient of refusing charity or public support to any families which could not support themselves. This grim prescription he justified on the ground that it was the only humanitarian remedy in its ultimate effect, shortsighted benevolence being a palliative which would aggravate the disease. He did not, however, advocate abolition of private charity to those who, being responsible and industrious, had suffered some undeserved calamity—the people who came to be known as "the deserving poor."

Malthus's doctrine served the immediate interest of those who, though profiting richly from the growth of capitalism, were under attack because of the miserable plight of a large proportion of the wage-earners. Human misfortune, he was arguing by implication, was due to disregard of a law of nature; no social benefit could arise under any economic order if this law was ignored. The remedy lay in the hands of the unfortunate themselves; if they were miserable it was their own fault. The only obligation of the upper classes was to instruct the people about the true situation. It was the views of Malthus that led economics to be called "the dismal science."

The logic of Malthus's analysis, if not of his remedy, has profoundly affected all social thinking. Since 1800 his predictions have failed to come true in the Western world only because of the very rapid increase in industrial and agricultural productivity, which he did not foresee, and some decline in population growth due to a falling birth rate. But nations like India and China still seem subject to the fate he outlined. It is of supreme importance to the more fortunate nations to devise some world-wide remedy; if possible a better one than those recommended by the Reverend Mr. Malthus.

Adam Smith's Followers

Like every great thinker, Adam Smith inspired disciples who did not bring to the subject much sense of reality or freshness of insight, but elaborated, refined, or corrected his theories, made a "school" out of the doctrine, and became more concerned with technical details than with broad issues and changing institutions. There is always a danger that important bodies of doctrine will degenerate into the sort of inconsequential logic-chopping which characterized the medieval schoolmen at their worst. In such circumstances not until a new approach to natural phenomena is possible will real advances be made.

In France, Jean Baptiste Say (1767-1832) rewrote Smith in a more systematic, logical, and self-contained system and shared his reputation. Say somewhat broadened the doctrine by contending that all producers of "utilities"—that is, things people want and are willing to pay for—are producers, whether they work on physical material or merely provide services.

He also propounded the statement, which came to be known as Say's law, that since the production of any article creates an equivalent demand for some other article, total supply must equal total demand, and so there can be no such thing as general overproduction. This observation was long taken to mean that the depressions that we all know occur could rise only from energy misdirected in producing more of one or more kinds of goods than were wanted. It obstructed acceptance of the type of explanation which seeks the clue to the business cycle in variations of the circulation of income or spending. Of course, there *can* be general overproduction in the sense of more goods than can be bought with existing income at current prices.

James Mill restated classical doctrine in oversimplified form with emphasis on its "laws." John Ramsay McCulloch was also a dogmatic though lucid economist who added little. Nassau William Senior, more eminent and able than these two, was a particularly good logician and typified the aridity of the school. He narrowed the scope of economics to exclude welfare or moral opinions, asserted that complete objectivity should prevent the economist from giving any advice, and assumed that the whole truth of the matter could be deduced from premises known to almost everybody. He was also particularly concerned with precise terminology. In all this he thought he was adopting a scientific attitude and held that economics had already attained the dignity of a science, since its basic laws had been discovered.

David Ricardo

Closely following Smith and Malthus in time, Ricardo (1772-1823) is the most celebrated classical authority except Smith himself, especially among professional economists. He is noted for the precision of his thinking, his power of abstraction, and the ruthless disinterestedness with which he followed his logic. His method was almost entirely deductive, omitting those salty observations and simple examples from real life which made Smith delightful to read. Compared with Smith, Ricardo is difficult and dry. Yet his ideas had momentous consequences.

Ricardo was descended from Spanish Jews and learned the business of stockbroking in his father's London firm. There he became familiar with the principles of banking, foreign exchange, and finance in the largest financial center of the world. By marrying a Quaker and becoming a convert to the Anglican Church, Ricardo displeased his father. He set up his own firm and soon became rich in his own right. At about thirty-five he was a multimillionaire. Then he indulged his liking for study, devoted himself to mathematics and science and after reading the *Wealth of Nations* concentrated on developing the theories of political economy.

In view of his celebrated theory of rent, it is interesting that Ricardo himself became a large landlord. He also was a member of the House of Commons.

Ricardo's Contributions to Theory

In *The Principles of Political Economy and Taxation*, published in 1817, Ricardo was concerned chiefly with the problem of the distribution of wealth, which he thought neither Smith nor Malthus had satisfactorily explained, though he took much from both of them.

Rent, Smith had declared, was a monopoly price. Ricardo agreed, but he elaborated the idea much further. If land were as abundant as air, it would be, like air, appropriated by anyone who wanted it and would command no price. It would be a "free good." This, he believed, was originally the case. The first farmers appropriated the best land. But as soon as the best land was exhausted, somebody would take over a tract that was not so fertile. The best land would thereupon command a price, since, acre for acre, it would yield more and there was no more of it to be had free. As the process continued, less and less fertile land being appropriated, value would accrue to all the better grades of land.

The least fertile land that was in cultivation would be farmed only if it would pay for the labor necessary to work it. It would yield just that and no more. The rent charged for the better land would therefore not be a payment for labor, but rather a payment resulting from mere ownership of a scarce form of natural resources. This payment, which Ricardo identified as rent, was unearned. Incidentally, in introducing the concept of the role of "marginal" land, he set forth an instrument of analysis later generalized in theories of value and prices.

Ricardo took over and systematized Malthus's iron law of wages. Wages could never be far from the level necessary to maintain a minimum of subsistence, because of the action of demand and supply in the labor market—higher pay would increase the labor supply, lower pay would decrease it. But he went further, to analyze the cost of subsistence. It would depend mainly on the price of food and other farm products. High crop prices necessitated higher wages. Higher prices of crops, in turn, were due to rent, which rose as a nation became more fully populated and the best land was exhausted.

In charging higher prices for the means of subsistence, landlords were exploiting, not labor, but the employer who had to pay the higher wages. The employer could not charge any more for his product because he had to pay the higher wages, since the prices he received were determined in competitive market.

The "natural price" of any article was based, as Adam Smith thought, on the cost of the labor that went into it, though Ricardo differed with Smith's opinion that rent was a cost that entered into the natural price. He did include in the natural price the cost of the labor required to construct the buildings and machinery, in other words, the capital. In receiving profit the capitalist was therefore taking something that labor had produced. Thus Ricardo left the wage-earner and the employer in conflict about division of the income of industry, and the employer and the landlord in conflict about division of the profit. Rent was in essence an encroachment on profit. The long-term tendency was for profits to fall to zero, while landlords would pocket the economic surplus without working for it.

Ricardo correctly saw that if the transactions of any nation with the whole world were taken into account, there must be an even balance between outgoing and incoming payments. He carried further and elaborated Mun's observations on foreign exchange. He also had much to say about money and the effect of differences in its quantity.

Implications of Ricardo's Doctrine

Ricardo, as an even more thoroughgoing disciple of laissez faire than Adam Smith, favored the complete abolition of protective tariffs. His doctrine of rent gave a particular urgency to the movement for repeal of the "Corn Laws" for protection of English agriculture—a change which not much later was made. Repeal of the Corn Laws was one of the influences which helped make Britain largely a manufacturing nation, emphasizing foreign trade and finance.

Also implicit in his doctrine of prices and wages, though he was probably unconscious of the conclusions that would be drawn from it, was the foundation of Karl Marx's theory about the exploitation of labor, as we shall see.

4

The Early Socialists; Karl Marx

We have been following the ideas of those thinkers in the eighteenth and early nineteenth centuries who interpreted liberty, the dignity of man, and the supremacy of natural law in a way that supported the emerging economic order—capitalism. Their stress fell on the self-seeking of the individual as a necessary and beneficent social force.

The optimism of the Physiocrats and Adam Smith had, however, undergone a sad decay in such doctrines as those of Malthus and Ricardo. In these theories labor could never for long receive more than a mere subsistence and might fall below that, no matter how much the sum of all wealth might be increased. Worker and capitalist were antagonists; landlords became the enemies of all other consumers. Wide disparity of incomes was accepted as natural and unavoidable; anything like economic or social equality was lacking even as a distant goal. Even philanthropy could only increase misery.

But another tradition appealed almost equally to men of good will, though it ran counter to the prevailing current of history. This emphasized the social aspects of the supposed natural order—readiness of men to cooperate, perfectability of human nature, and the desire for equality not only political but social and economic as well. This type of philosophy, though often condemned as materialistic, laid no more stress on material goods than the doctrines of the individualists. Its leading exponents were concerned with economic arrangements only as an instrument to open the way for what they regarded as the natural nobility in all men, which they supposed had been thwarted by maltreatment, misery, and miseducation.

53

The roots of egalitarianism lay far back of the Enlighten ment. The demand for equality had raised its head in ancien Greece; it had been stimulated by the Christian teaching o the brotherhood of man. During the English Puritan revolution of the seventeenth century the "Levellers" tried to swing the balance of power not only from king to commoner, but fur ther, to the propertyless worker. Gerrard Winstanley, leader o the Diggers, assumed that property had been held in common until the Norman Conquest and asked Cromwell to restor free land. "As every one works to advance the common stock so every one shall have free use of any commodity in the store house for his pleasure and comfortable livelihood," he wrote "without buying or selling or restraint from anybody." It wa an idea bound to occur to many, in many times and ages.

The French Utopians

The Enlightenment and the French Revolution led to elabo ration of this kind of thinking. Rousseau had contended tha private property was robbery and had not existed in the stat of nature. The French Revolution, however, was not so muc the child of the Paris mobs as of men of comparative wealt and standing, and though from time to time it threatened t get out of hand, its essential change turned out to be victor of the business classes over the landed nobility. After it, th plight of the common people was scarcely relieved, since th growth of early industrialism in France, as elsewhere i Europe, was accompanied by unemployment, long hours, sta vation wages, and the decline of agriculture. The resultin disillusionment created a favorable atmosphere for the apostle of economic equality.

François Émile Babeuf (1760-1797), himself a product c the Revolution and a supporter of the Reign of Terror, co spired to overthrow the Directory and establish a communist society. He called himself "Gracchus," after the idealistic lead er of the Roman proletariat. The conspiracy was discovere and Babeuf was condemned to the guillotine. He held th "nature has given to every man an equal right in the enjoymen of all goods." There was to be immediate national ownership all large business enterprises, and eventual nationalizatio of all private property by abolishing inheritance. Productic and distribution were to be directed by an elected governmen Nobody could have political rights who did not do useful wor and teachings contrary to the regime were to be forbidde Food and clothing were to be exactly the same for all, exce or differences according to sex and age. Children were to be taken from the parents and taught the ways of the new society.

This rigorous doctrine was echoed in somewhat milder terms by Étienne Cabet, who wrote a romance, *The Voyage to Icaria*, instead of conspiring to cause a violent revolution. Cabet was born in 1788, during the Revolution, studied law, and became for a while a public official. Icaria, Cabet's hero reports, was a sort of technical dictatorship in which uniformity of every kind prevailed. The streets were straight, and each block contained exactly fifteen houses just alike, with the most modern sanitary equipment. Glass roofs covered the sidewalks, and dust-collecting machines swept the streets. The state owned everything and divided the product equally. Everybody dressed alike, though personal taste was indulged in color. No newspapers were allowed, and books had to be submitted for government approval before publication. The whole program was carefully laid out according to the decimal system.

Cabet aspired to create such a regime within fifty years and believed that it could be approached by easy legislative stages. He emigrated to the United States and obtained a grant of land in Texas, where he set up colonies which he hoped would generate imitation. Yellow fever rudely interrupted his plans. Then he moved to Nauvoo, Illinois, where a colony of fifteen hundred, presumably without glass-covered sidewalks, was broken up by internal quarrels.

Saint-Simon

Comte Henri de Saint-Simon (1760–1825), a Parisian, was a less arid but equally optimistic thinker. His plea was not for uniformity but for equality of opportunity. He fought in the American Revolution and supported the Revolutionists in France. Thereafter he endured hunger and cold and sacrificed his health in a long struggle to develop his ideas and write books which would so appeal both to reason and to ardor that their publication would soon be followed by a golden age. The chief of these books is *The New Christianity*.

Men in the new order, he contended, needed a new spiritual authority which would play the role of the Church under feudalism. The new order, which had been made possible by the destruction of the old, was to be scientific and industrial. Destruction of the old, though necessary, was not enough; it must be followed by something better than anarchic individualism. War must be eliminated; Europe must be united under

a single parliament through which the wise and the just woul
govern, their minds being illuminated by science. "In the Ne
Christianity, all morality will be derived immediately from th
principle: men ought to regard each other as brothers."

Such a regime would, Saint-Simon believed, install publi
ownership of industry but leave consumption goods in priva
hands. Each should receive an income corresponding to h
services. The delicate problem of apportioning the paymer
for work according to merit was to be left to public official
No idlers, rich or poor, were to be tolerated.

Saint-Simon had many followers among intellectuals an
scientists, including Auguste Comte, the philosopher, and Fe
dinand de Lesseps, the engineer who built the Suez Canal.

Fourier

Charles Fourier (1772–1837), son of a merchant, hoped t
convert the world to a better system by example rather tha
by preaching. Associations of cooperators, or "phalanxes
were to demonstrate the superiority of communal living.

Fourier as a boy had been deeply depressed by the corrup
tion of the commercial world when his father punished hi
for telling a customer the truth about a product; this feelin
was strengthened later when he was ordered to throw ric
overboard from a ship because the owner, speculating o
higher prices during a famine, had kept it until it spoiled

Fourier had some fantastic notions, such as that the worl
was about to enter a millennium in which lions would dra
carriages, whales pull ships, and sea water be drinkable. Th
sort of thing may have been intended as symbolism, or eve
as publicity, but it had little to do with his plan for th
phalanx. That involved a small colony living in a commo
building, apportioning the tasks of sustaining the group accor
ing to personal taste, and thus, through the joy of work for th
common good, combined with efficient management, increa
ing output enough so that everyone could retire in comfo
at the age of twenty-eight.

The sanguine author announced that he would wait eve
noon for a rich man to call and offer to finance such a colon
but though he was at home at the appointed hour every day f
twelve years, none appeared. The proposal, however, obtaine
wider approval after his death. A few phalansteries were tri
in France, and in the United States the idea of "associatio
ism," as it was called, was picked up by pre-Civil War refor

rs revolted by the spirit of private gain and exploitation which prevailed under the newly fledged industrialism. Among the converts were Albert Brisbane, Horace Greeley, and Charles A. Dana. One of the colonies tried was that at Brook Farm, in which Margaret Fuller, Nathaniel Hawthorne, Bronson Alcott, and other noted Americans took an interest. None succeeded, though some endured for a long time; this experiment, like many in the chemical or physical laboratory, failed to produce the desired result. The failure of utopian colonies was used to ridicule belief in the possibility of a better society. It is easy to make fun of the utopians, but the influence of their ideas cannot be measured by the crudeness of most of their schemes.

Blanc and Proudhon

Laissez faire was rejected also by Louis Blanc (1811–1882), who was distinguished as the first socialist who appealed to the workers themselves to bring about reforms and who used the instrument of the state in an effort to create the new society.

Son of an inspector-general of finance in Spain appointed by Louis Bonaparte, Blanc became a journalist and at twenty-eight founded the influential *Revue de Progrès*, in which he published serially in 1840 his chief work, *Organisation du Travail* (*Organization of Labor*). During the Revolution of 1848 he became a member of the provisional government. Forced to leave France during the reign of Napoleon III, he returned in 1870 and was elected to the national assembly, but opposed the insurrection which tried to establish the Paris Commune.

He proposed a sort of WPA—social workshops established by the state where everyone could be guaranteed a job—but, unlike the WPA, these were to be not a form of emergency relief but permanent institutions to constitute the basis of a new society. He believed that through efficient competition with private enterprise they would soon drive other businesses out of existence. The workshops, under the workers' control, were to form a national federation to insure against the losses that might be incurred by any establishment.

Blanc did not believe that all men were equal in talents and enunciated the famous socialist formula, "From each according to his ability, to each according to his needs." This did not imply equality either in tasks or in distribution, since needs as well as talents differ. The whole plan was to offer an opportunity for human development denied by the competitive system of *bellum omnia contra omnes* (war of all against all). In 1848 the plan for social workshops was tried, but under an enemy of

Blanc, for the express purpose of discrediting him by its pre
dicted failure. The life of the workshops was short.

Pierre Joseph Proudhon (1809–1865), though a believer in
equality and a bitter enemy of private property in business, dif
fered from the socialists by opposing the state as well. He
visualized the ideal society as one which would embody "the
union of order with anarchy." Without government private
wealth could not exist, since there would be nothing to protect
the owner. Those who had first broken up communal owner
ship by appropriating land were essentially thieves. Employers
robbed labor by not rendering to them the full value of their
labor. The basic principle of society was that everyone was en
titled to the product of his own work; this kind of property
was natural and would be respected by all if the state no
longer existed to protect those who stole from the community
and exploited others.

The doctrine of anarchism has been influential in the Euro
pean labor movement, particularly in the syndicalist unions of
France and Spain, and was for a time embodied in the philoso
phy of the American IWW. Strong traces of it can be found in
Karl Marx, though he is not regarded as an anarchist.

Jeremy Bentham's Utilitarianism

An English social philosopher, Jeremy Bentham (1748–
1832), though like Adam Smith a believer in laissez faire as
a policy, introduced a point of view important in the develop
ment of socialist thought. Son of a wealthy solicitor, he studied
law, but he did not have to earn a living and decided to
devote his life to the promotion of human welfare.

"The greatest good to the greatest number" was his an
nounced aim. Every social institution was to be judged by its
usefulness in increasing the good to individuals—hence the
name attached to his doctrine, utilitarianism.

How was good to be recognized? Anything that increased
the pleasure or diminished the pain of any person was good.
Social good was the algebraic sum of all individual goods.

In order to give concrete meaning to this aim, one had to
have a means of measuring pleasure and pain in millions of
different individuals. The common and convenient measure,
Bentham believed, was money. Lack of money was responsible
for misery; enough money could bring happiness. But if the
wealth of an individual exceeded a sufficiency, his pleasure
did not increase proportionally. The same amount of wealth
if available to the poor, would bring a much larger sum of

appiness. Thus Bentham's logic led inevitably toward the
esirability of greater equality of wealth, a policy which was,
n one form or another, also that of the socialists.

Bentham made a contribution, too, in rejecting Adam
mith's belief in the beneficence of nature. "Natural rights"
e thought a mystical absurdity. In order to maximize utility,
t was necessary to employ reason to discover the best policy
or the end in view. Economics was not merely a science, de-
oted to analyzing what existed, it was also an art in the
haping of human affairs. Man could not rely on a personified
Nature to do this for him.

Though Bentham's reasoning led him to endorse quietism
y the state as the best policy, it could lead others to advocate
tate intervention. Indeed, Bentham himself was prominent
mong the philosophical "radicals" who promoted British
eforms of the period. Among these reforms were not only
hose which repealed governmental interferences, such as pro-
ibition of organization of laborers, protective duties, and
iscrimination against Catholics and Dissenters, but those
vhich activated the state, such as extension of popular educaion and of the right to vote, public sanitation and hygiene, a
ew poor law, the famous Reform Bill of 1832, and the
Municipal Corporations Act of 1835.

Robert Owen

It is a curious but significant fact that the man who had
nore to do than any other with inspiring the British socialist
novement in its early stages was a highly successful manufac-
urer—Robert Owen (1771–1858). Born in North Wales into
he family of a small businessman, he quit school at the age of
ine, continued his own education by reading, and at nineteen
orrowed £500 from his father to set up his own business as
 cotton spinner in Manchester. He was the first British mill-
wner to use the American long-staple cotton grown on the sea
lands off the Carolina coast, and became highly successful.
efore he was thirty he had purchased the New Lanark Mills
ear Glasgow, employing about two thousand persons.

Owen was a good manager. He made ample profits but was
nore interested in doing a thorough job than in accumulating
ealth. He used the latest methods and machinery, which he
naintained with scrupulous care. His mill was orderly and
potless. It seemed to him ridiculous to pay so much attention
o physical equipment without being equally thrifty about the
uman lives engaged in the undertaking. The prevailing situa-

tion of mill operatives shocked him. Many were women and
children, poorly clothed and housed, underfed, and worked
until exhausted. Man of action that he was, he startled everyone by doing something about it.

He reduced hours and raised wages. He built model hous-
ing, installed free education, and placed in the schools all children under ten, whom he no longer would employ. Children
above that age were instructed on working time. In order to
provide nourishing food and decent clothing at low prices he
opened a company store. Fines for spoiled work were abandoned. Recreation was provided, and insurance funds set up.

Nobody else had ever thought of doing such things. It was
hard enough for Owen to persuade his partners to consent to
the expense, though the business did not seem to suffer. He
even carried them along when, with the mills shut down for
four months because of depression, he paid the employees full
wages. Owen was pleased because conditions in the town had
markedly improved. No longer were poverty, disease, and
hopelessness prevalent; it was an orderly, clean, and pleasant
community. Distinguished visitors came from far and wide to
see the miracle. But industry as a whole failed to follow the
example.

Owen had been deeply impressed by the idea that human
life could be improved by a better environment; his own experience confirmed his conviction. Now he was further impressed by the futility of depending on paternalistic employers to change conditions. They did not have the will, and he
came to the conclusion that they did not have the power either
since the addition of profit to wages and other necessary costs
made prices so high that the workers could not buy what they
had produced. Only a cooperative system, without capitalists,
could be humane. Characteristically he began at once to work
toward fulfillment of his idea in his own establishment. As a
step toward the new system, he decided to limit profit to five
per cent. That was at last too much for his partners. The firm
was dissolved. But he succeeded in finding new partners, among
them none other than the celebrated Jeremy Bentham.

The deep and world-wide depression following the defeat
of Napoleon in 1815 then convinced him that an entirely new
start had to be made, and so in 1825 he founded two cooperative communities, one in Scotland, and the other in New Harmony, Indiana. Neither worked out well, and they consumed
most of Owen's wealth. Undaunted, he tried something else in
1832, the National Equitable Labor Exchange, in which anyone could deposit the products of his labor, receiving in exchange notes with a purchasing power equal to the number of

ours of work spent on them. The notes could be used to buy he deposited products. Thus there could be no difference in otal value between demand and supply. But the articles for ale did not attract enough purchasers; apparently they were ot regarded as embodying full value even though their price ontained no employers' profit.

Owen helped to organize unions. He never ceased preachng the virtues of cooperation, the creative influence on charcter of a better environment, the need for universal educaion. The great consumers' cooperative movement, springing rom a small beginning in Rochdale, England, owes much to is inspiration, as does the British labor movement. His writngs are not intellectually profound, but their eloquent expression of faith in the plastic nature of humanity, as in *A New View of Society*, has acted as a tremendous influence. British socialism probably owes more to the tradition he embodied than to that of Karl Marx.

"Any general character," Owen wrote, "from the best to he worst, from the most ignorant to the most enlightened, nay be given to any community, even to the world at large, y the application of proper means; which means are to a great xtent under the control of those who have influence in the ffairs of men."

Karl Marx

The great prophet of modern socialism was Karl Marx 1818-1883), who combined the ideas of many of his predeessors, both socialist and capitalist, with others of his own, o form an ambitious new system of thought which has beome the basis of a powerful movement as the years have assed, despite the flaws in his thinking and the errors which istory has demonstrated in his predictions.

Marx was at once a philosopher, a historian, a sociologist, n economist, and an active controversialist in the struggles which characterized his lifetime. He outlined a framework of he future course of events on which he based prescriptions or strategy on the part of those who wished to change the ature of society.

So extensive and difficult are his written works, and so comrehensive is his view, that he has suffered the fate of most najor prophets. By lesser followers his doctrine has been ransmuted into dogma, citations of his works have been subtituted for fresh observation and analysis, factions and schisms ave arisen among the "true believers."

Marx's Life

Marx was born in Treves of German-Jewish parents who had become converted to Christianity. He studied in German universities, obtaining his doctorate of philosophy while under the influence of the leading German philosopher of the day Friedrich Hegel. Democratic views which he early developed thwarted his ambition for a university career, and thereafter he earned a precarious living as a journalist. The autocratic Prussian government suppressed the *Rheinische Zeitung* of Cologne in 1843 because of his articles. Then, having recently married Jenny von Westphalen, a girl of aristocratic family who had been his childhood playmate, he moved to Paris.

In Paris, Marx met Proudhon and other anarchist and socialist leaders, with whom he had long and earnest discussions, in most of which he violently disagreed with them about doctrines and methods. He also made the acquaintance of his lifelong friend, patron, and collaborator, Friedrich Engels, son of a millowner with interests in Manchester. Marx's journalistic writings in Paris, dealing largely with the German situation displeased the Prussian government, which extended its long arm to bring about his expulsion from France. He then settled in Belgium.

The 1840s were years of turbulence, culminating in the revolutionary movement of 1848, notably in France, where the monarchy had been restored after Napoleon's defeat, and Germany, where in the monarchical states democracy had never been introduced. The industrial workers, subject to repeated crises of unemployment, low wages, long hours, and deplorable working conditions, were particularly rebellious though dissatisfaction permeated other levels of society as well. A secret international organization, calling itself the League of the Just, thought in 1848 that the opportunity of exploited labor had arrived and decided to proclaim its views and head the revolutionary movement. They asked Marx and Engels to draft their proclamation, and the result was the celebrated *Communist Manifesto*.

This document asserted the existence of a class struggle between capitalists and workers, stated that everybody would be liberated with a final victory of the working class when it took over the means of production, and ended with the slogan "The proletarians have nothing to lose but their chains. They have a world to win. Workingmen of all countries, unite!"

Anyone now living who is stirred by these ringing words

may be surprised to discover, when he reads the full text of the manifesto, that most of the benefits promised labor as a result of seizure of power have long since been achieved in so-called capitalist nations—for it announced such goals as the eight-hour day and social insurance. Modern opponents of these reforms have often called them socialistic on the ground that they were included in the demands of socialists. But nearly all workers spontaneously desired shorter hours and greater security, whatever their political leanings, and in 1848 Marx and his colleagues firmly believed such gains never could be achieved under a capitalist regime. Experience at the time seemed to support this conclusion.

After the collapse of the revolutionary movements of 1848, Marx moved to London and spent the rest of his life elaborating and supporting the doctrines foreshadowed in the *Communist Manifesto*. He lived meagerly, supported in part by a yearly contribution from Engels, and in part by acting as European correspondent of Horace Greeley's *New York Tribune*. The first volume of *Das Kapital* (*Capital*) was published in 1867. The second and third volumes were published by Engels in 1885 and 1895 respectively, after Marx had died. The first volume contains the gist of his teachings.

Marx's Economics

The economic theory developed in *Capital* is almost wholly classical, much though the discovery may surprise both orthodox followers of Smith and Ricardo as well as orthodox socialists. Marx used no assumption not outlined by some writer of the classical school, and his method of reasoning was, like theirs, deduction from a few relatively simple postulates. If his theory is mistaken, as it almost certainly is, so is theirs, and if this method is prone to lead to insupportable conclusions, so does theirs. He was, to all intents and purposes, a classical economist, the only difference being that in his case the reasoning became a weapon to attack capitalism rather than a weapon to defend it.

Like most of the writers who preceded him, Marx attributed the economic value of commodities to the labor expended on them. This was in essence Ricardo's view. Marx added merely that in order to produce value the labor must be socially necessary, that is, its product must be of use to somebody. Value in exchange, as markets develop, may be measured in money terms, but if all illusions and deceptions of money are penetrated, the consumer who wishes to use an article is mere-

ly paying for the socially necessary labor that went into i
Money is the intermediary.

The capitalist, however, is not buying articles for use, h
is buying articles for resale. A cotton-mill owner buys yarn
spindles, and the labor of the spinning operatives. Each c
these components of the final product costs him the price c
the labor that went into it. But he manages to sell the produc
of his mill for more than it costs him. If he did not, ther
would be no profit in the transaction. Therefore he is the re
cipient of *surplus value*.

What the employer has to pay the worker is only enough t
keep the worker alive. (Here is the Malthus-Ricardo "iron la
of wages.") What he collects from the customer is the tru
value of the labor put into the article. The capitalist, in th
process of accumulating wealth for himself, therefore robs an
exploits the worker.

Let us assume, for example, that it would take only a half
day's work to produce the minimum of subsistence of a wage
earner for a whole day. Since the employer need offer no mor
than is required to maintain the supply of labor, he will pa
a wage equivalent to the minimum of subsistence, or for
half-day's work. But does he require his employee to wor
only half a day? By no means; he insists that the worker sha
put in, not six hours, but twelve.

Or suppose the employer introduces new machinery tha
will double the output of the worker in a day. Does he
double the wages? Not at all; he keeps the surplus value fc
himself. Only in such ways is privately owned capital ac
cumulated, since if the capitalist paid out for labor all th
money he received, he could accumulate nothing.

This theory, it should be noted, does not imply that th
world needs no so-called "capital goods" such as machiner
or buildings. The purchase of such goods for purposes c
production would be a necessary part of any social system
it would, in essence, be payment for the labor that produce
the goods. What the theory implies is that the private em
ployer has no right to claim such goods as his own. And thi
was just the conclusion Marx was striving to justify. If th
workers should own the instruments of production, they woul
presumably set aside part of their earnings to pay for ne
machinery and buildings.

In thus putting together the classical theory of value an
the classical theory of wages to demonstrate the injustice o
capitalism, Marx was preceded by Johann Karl Rodbertu
(1805-1875), a German economist. Rodbertus also foreshad

owed Marx in his theory of economic crises. By failing to distribute to labor the full value of its product, Marx thought, the capitalist continually tended to produce more than the market for his goods would absorb. Products would not automatically be "cleared" as in a natural or barter economy. This was the cause of the frequently recurring periods in which a surplus of goods in the hands of the capitalist was accompanied by unemployment and want on the part of the workers. Though Marx's theory of depression was a crude and incomplete one, he deserves credit as one of the first economic theorists to emphasize an obviously critical defect in the operation of a capitalist order.

Marx's Philosophy of History

Hegel's abstruse philosophy, dealing with the realm of ideas, was taken over by Marx and applied to the development of society. A major tenet of Hegel's was that change takes place through a so-called process of "dialectics." For every positive there is a negative; for example, white and black, good and evil, high and low. Ideas, beliefs, systems of thought are arranged in opposite pairs. Every positive Hegel called a "thesis," its negative an "antithesis." The opposition of the two could lead to a new concept, called a "synthesis." The synthesis in its turn would become a thesis, with its appropriate antithesis.

Marx applied this doctrine to account for a process of social evolution based on man's concern with material things, the doctrine known as "dialectical materialism."

Like other thinkers of his time, he supposed that there had been an original primitive society in which equality and cooperation had prevailed. But, he observed, quarrels had arisen about power and wealth; history was a long record of gradually intensifying conflicts in which one group or class tried to exploit another. Some of his most eloquent passages describe oppressions, injustices, and miseries which had arisen in this way. How and when do such conflicts arise, how do they end, and what is the result? To these problems Marx applied his ingenious theory, the logic of which was derived from Hegel.

A society like the primitive one, which was in social equilibrium, was disturbed because new means of doing things were invented. New weapons, new tools, new methods began to be applied. But the old ways of thinking and codes of morality were not adapted to the new situation. Like all systems of thought, they were derived from the material situation in which they arose. Ideas do not create social systems; social systems generate ideas.

A new "superstructure" of institutions and beliefs is required. The persons who profit from the old regime resist change. A class conflict arises between the conservatives and the sufferers from the misfit between the new techniques of production and the old order. When this conflict becomes sharp enough, the collision between thesis and antithesis produces a synthesis—the new social system. This abstraction is a reasonably informative though oversimplified explanation of the process by which feudalism gave way to capitalism, as Marx demonstrated.

Now, he observed, a new class conflict was coming to a head within the capitalist order—the conflict between capitalists and exploited wage-workers. The capitalists had had their necessary role to play in introducing more highly productive techniques of division of labor, factory production, machinery, and world-wide trade, against the opposition of the feudal interests. But the system was operating to the disadvantage of those who performed the labor. Factories and machinery could not and should not be abolished, but the rules by which the economy was governed needed changing. The ruling classes would be deposed and new rules would be adopted when, in the course of social evolution, the time was ripe.

This idea of social evolution Marx believed "scientific," since it was arrived at by a process of observation and deduction just as Darwin had arrived at the idea of biological evolution. It caused the Marxist doctrine to be called scientific socialism as opposed to the schemes of the utopians, reformers, or advocates of changing everything overnight by conspiratorial violence.

Marx had no faith in idealistic colonies and patent-medicine schemes which ignored the major tides of historical development. He thought that it was useless to overthrow a government before the circumstances were propitious for genuine change. He had no confidence in the success of benevolent individuals whose action, if it were to be effective, must be in accord with the institutions of the system that hemmed them in, the system that itself was responsible for injustice. The system could be abolished in due time by the inevitable opposition of those who suffered from it—the working class—and in no other way. Capitalism would not give way before it had reached its highest stage of development and had revealed clearly, by its own operation, its incapacity to serve human satisfactions.

What socialist leaders could do was to point out to rebellious workers the real causes of their misery, win their confidence by supporting their struggles, and, at the appropriate moment of crisis, be prepared to show them what to do. The natural laws of social evolution were as inexorable as Newton's laws of motion, but as in the case of laws of physics anyone who understood them could be prepared to utilize them to human advantage.

Marx's Predictions

One recognized test of any scientific theory is the ability to make valid predictions by use of its logic. Marx did not hesitate to apply his theory in this way.

The natural laws of capitalist development, Marx believed, would continue to lead, as so far they did seem to be leading, to the elimination of neutrals in the class struggle, so that a relatively few capitalists on one hand would confront a propertyless working class, or "proletariat," on the other. The proletariat would comprise all but a tiny minority of the population. It could come into power either by exercising democratic rights to vote where such rights were universal and not subject to corruption, or by violent struggle where the ruling class managed to thwart the will of the majority. Marx did not have much faith in the peaceful surrender of the owners, though he admitted the possibility that it might occur in nations like Britain and the United States.

This division of the people into two opposing forces would come about through the spread of industry and big business, which would gradually absorb farmers from the land, take away the occupations of handicraftsmen and small tradesmen, and concentrate the ownership of capital among fewer and fewer persons. Farming, he thought, would itself become industrialized and subject to concentrated ownership.

Another prediction by Marx was that the working class would be subject to increasing misery as the development of capitalism approached its climax. Capitalism would be obliged, he thought, to keep wages low and hours long, since it could not find markets for the output of its steadily increasing productive capacity and must therefore extract the utmost possible "surplus value" from the workers in order to keep from going bankrupt.

In the end, capitalism would falter and collapse as a solvent system from its own "internal contradictions." It could maintain itself only by continual extension into new sectors of production in the countries where it was strongest, and into new areas of the world. Eventually it would have spread its blight so widely that there would be no more territory for it to conquer and exploit. In the meantime it would be subject to more and more severe periodic crises, caused by glutting of markets and accompanied by general unemployment.

Marx furthermore expected that the final crisis of capitalism would arrive in the relatively near future, perhaps during his own lifetime. If his analysis of the situation were correct, it could hardly be otherwise.

And after the Revolution?

It is curious that Marx's doctrine, unlike those of the utopians, contained almost no plans for the society which would arise after the workers took power. He seemed to assume that on the heels of that event all wrongs would automatically be righted, universal happiness would prevail, and his ponderous machinery of social evolution would grind to a stop. No new antithesis was suggested for the new thesis which would come into being with the fall of capitalism. Followers of Marx who look into his writings for guidance about running an established socialized society find almost nothing. After all, he seemed to accept the major premises of the utopians whom he inevitably excoriated. Only abolish private profit, and the natural or primitive state of cooperation and equality would automatically return, one is left to infer.

Marx did have something to say about the state—that is the organ of political government. His quarrels with the anarchists and his concern with the strategy of revolution made some view of the state inevitable. The state, he believed, was "the executive committee of the ruling class"; national sovereignty was merely a veil for class dictatorship. During the period in which he formed his opinions, a period characterized by limited suffrage, rudimentary popular education, suppression of labor organization, clearly visible class lines, and the unchecked supremacy of wealth, this epithet was hardly an exaggeration.

When the workers seized power, Marx thought, they would of course exercise it partly through the state; this would be necessary and desirable in the transition. Hence the phrase "dictatorship of the proletariat." But of course Marx had in mind a proletariat which would constitute almost all the population in a highly industrialized nation, not a minority, and still less an exclusive and dominating party. Once socialism was firmly established, however, Marx believed that the state would "wither away," since it would have lost its chief function—that of policeman for private property in the means of production.

You will scarcely find in Marx the word "planning," or any discussion of socialist distribution of income, or of money and prices under socialism, or of how a socialist society would carry on the process of investment, allocate its resources, avoid inflation, or conduct foreign trade. There is no hint that these and other problems might require an economic theory and policy of their own. And naturally so, since Marx's economic theory was centered on the attack on capitalism, not on economic policy *per se*.

Agricultural proprietors in their resistance to capitalism had found a rationalization in Physiocracy; capitalists had discovered their defense in classical economic doctrines; now a third great group, the industrial workers, were provided with a powerful philosophy of opposition to the society in which they, for the time being, were so unhappily situated.

Assessment of Marxism

A favorite and obvious method of undermining the conclusions of Karl Marx is to demonstrate the historical errors in his predictions.

Agriculture has not become subject to concentrated industrial ownership; independent businessmen and self-employed persons greatly outnumber big capitalists; the "industrial proletariat" proper, far from being almost all-inclusive, is not even a majority in most countries. The working class has not suffered increasing misery, but has enjoyed, in most industrial nations, a rising trend of real wages and higher standards of living ever since the middle of the nineteenth century.

There have been severe depressions and unemployment, but no proof exists that depressions are, on the average, deeper or longer than when Marx wrote. The proletariat has not carried out a socialist revolution in any highly industrial state; what now goes by the name of communism has won its victories chiefly in backward and marginal regions. Even the moderate socialistic measures installed in Britain after World War II by the Labour government were made possible by a slim electoral majority achieved by attracting middle-class voters.

It may be that Marx's predictions were wrong in timing rather than in substance. Perhaps a catastrophic collapse of capitalism lies somewhere in the future. Yet if it were as sinister and inflexible an order as it seemed to Marx at the time, it could scarcely have yielded all the benefits that have been extracted from it during the past century.

But it will not do to dismiss Marx for his errors. No great thinker was free of mistakes, or enunciated nothing but doctrines valid for all time. Taken as an innovator and a scholar, Marx made important contributions, though not so much to economics as to the theory of social development. His stimulus is part of the heritage of modern thought. Who can tell whether capitalism would have done as well as it has if it had not been so sharply challenged by persons and movements influenced, at least in part, by this type of thinking? Marx's doctrine, swallowed whole and used as an instrument of propaganda, can do great damage, but so can any body of doctrine treated as a universal and final expression of truth. If the ideas of Malthus and Ricardo had been applied to practical affairs as rigidly as some of their present-day adherents seem to wish, what we know as capitalism would probably have been overthrown years ago.

5

Other Economic Protestants

The rise of socialist thought in the very years when industrial capitalism was beginning to triumph over older forms of economic organization is a symptom that many suffered in the process and many sensitive minds were revolted by it.

Yet most socialists, both utopian and "scientific," accepted much the same basic assumptions and methods as the theorists of the classical school who supported capitalism. They regarded wealth as composed solely of material things. They posited a "natural" cooperative society somewhere in the past; they believed in natural rights. Their process of thinking was mainly use of deductive logic. Marx even founded his economic theory on the classical doctrines of value and wages.

Other germinal thinkers, equally outraged by capitalism as it existed, adopted intellectually more radical positions than that of the socialists by challenging the classical assumptions and methods themselves. They foreshadowed future broadening of economic thinking, not merely by turning Smith, Ricardo, and Malthus inside out, but also by introducing considerations which lay outside the focus of the classical view.

Sismondi, Herald of Reform

One of the earliest and most discerning critics of the classical school was Jean Charles Léonard Simonde de Sismondi (1773-1842). His aristocratic Italian family, originating in Pisa, moved to Geneva, Switzerland, where Sismondi was born. He pursued his studies there, worked in business in Lyon, and then went to Italy, where he devoted his life to scholarship and writing. Sismondi was noted chiefly as a historian, and it was against the background of his work in history that he projected his views on economics. His major economic study, *New Principles of Political Economy*, was published in 1819.

Sismondi asserted that economics should use as its criterion of values and its aim of policy not mere material wealth, as the classicists used the term, but human well-being. "Wealth is a modification of the state of man," he asserted; "it is only by referring to man that we can form a clear idea of it." It was necessary to know whether "man himself belongs to wealth, or wealth belongs to man." Wealth may be regarded as moral and physical enjoyment.

And we are concerned not only with how big the pile of wealth is. One must also ask, "For whom?" "No nation can be considered as prosperous, if the condition of the poor, who form a part of it, is not secure. . . ." In these attitudes Sismondi was the precursor of that eminent school of thinkers whose subject of study is now known as "welfare economics." His ideas led toward the modern type of society often called, both by those who approve and those who condemn it, "the welfare state."

He did not doubt that the classical writers correctly described the processes of production and distribution, but he did flatly disagree with Adam Smith that when men pursued their self-interest in making money, the consequence of their efforts, as if guided by an "unseen hand," was social benefit. Laissez faire he rejected as a basic principle. The state should intervene to protect human values. Though not against private property and not a believer in complete equality of income, he did think that concentration of wealth on one hand, and extreme poverty on the other, which would result from laissez faire, should be prevented.

In addition to promulgating this type of social policy, Sismondi was probably the first to make an extremely fertile suggestion about methods of economic thought. The deductive logic of the classical school, he pointed out, could lead to mistaken conclusions. Its abstract hypotheses needed to be tested by studies of the concrete facts. For that purpose he recommended reference to human experience—that is, history.

Two specific observations by Sismondi pointed out problems that have become a major concern of economics, though his analysis of these problems and his suggested remedies are open to serious question. The first was the observation that rapid introduction of machinery leads to technological unemployment and pauperism—as it actually did when he wrote. In consequence he recommended a dubious reform: limiting invention. The second was the obvious fact that severe recurring crises of unemployment beset the new order. This he mistakenly attributed to overproduction, in the sense that as production increases, a new year's output could not be bought with the preceding year's income. But he was far ahead of his time in concerning himself with a major economic phenomenon to which the classical school, in its clouds of abstraction, paid little attention.

Sismondi's immediate influence was small, but as a prophetic voice he has turned out to be, so far as the economics of the Western world are concerned, far more nearly justified than Karl Marx.

Müller, Apostle of Social Community

Just six years after Sismondi's birth, Adam Müller (1779-1829) was born into a German Protestant family in Berlin. Strongly influenced by the German philosophical and literary movement usually called Romantic, he applied it to political economy in a reaction against Adam Smith and the British classical economists even more profound than that of Sismondi.

Romanticism was a mystical and inchoate movement which emphasized the community of souls and the merging of the individual into a larger whole. It was capable of being employed to exalt the state as an organism with a life of its own, superior to individual rights and individual welfare. This strain in the history of German thought has often been condemned by writers in the Western democracies; it can easily be traced, for example, in the belief of many Germans in the virtue of military conquest, and was directed to evil purposes by the Nazi regime. But in Müller's hands it was significant as an antidote to Adam Smith's extreme economic individualism.

Mutual interdependence and the wholeness of life was Müller's dominant theme. It was absurd to try to isolate, even in theory, wealth-gaining activities from art, religion, or the services of the state. All were productive of utilities to human beings. Capitalism, and the theory of the classicists which justified it, led to a disruptive atomization of society. Capitalism created a deplorable breach between capitalists and laborers by removing the tools from the ownership of the man who worked with them. "The spirit reacts unceasingly against the division and mechanization of labor, which Adam Smith prized so highly; the spirit wants to preserve man's personality."

Society should be an organic whole, not a congeries of

warring individuals. Value does not inhere in material things only, and is not derived just from their usefulness to individuals. There is such a thing as "sociality" value too. "If we say of a thing that it is useful, we mean that it has a value in relation to civil society." Not only commodities had value, but "the words of the statesman, which would perhaps bring millions of actual money into existence; the words of the priest or those of the artist, which might ennoble the heart or enlarge the imaginative faculty of the nation."

Money has value not as a mere commodity, Müller asserted but is of importance because of its social function. Money is money only "insofar as it is not private property but is the common property of as many persons as possible, and, indeed of all. . . . Only at the moment of exchange, or of the circulation of the substances of money, are these latter really money."

Though his thought lacks precision and thoroughness of detail, Müller's writings contain a hint of the contemporary emphasis on the importance of the circulation of income and the social usefulness of a managed supply of paper money.

The factors of production include, Müller asserted, not only land, labor, and capital, as the classicists thought, but also "spiritual capital." In other words, a nation can have the material elements of prosperity without knowing how to combine them in such a way as to increase the values of living.

Such a combination of vague general ideas could be used either for progressive or for reactionary purposes. Müller in effect took the side of a return to the past. He wanted to go back to something like feudalism, though he criticized feudalism because its union of social forces "was effected federally rather than organically." In the end he became an official for Metternich, the Austrian statesman who united Europe for a while in protection of the old order against the waves of modernism liberated by the French Revolution.

Yet Müller's frontal attack on classical economics has found many an echo in more recent times. It has something to say about the spiritual malaise which many feel in a mechanized and materialistic civilization; it expresses a widespread desire to mold it into something nearer to the heart's desire.

William Morris (1834-1896), the British socialist, later emphasized this theme when he extolled the virtue of handicrafts both for the sake of the craftsman and for the sake of honest and beautiful products. The British Guild Socialists, about the time of World War I, argued that the industrial worker as a human being was degraded because in machine industries he was condemned to routine repetitive processes instead of being able to enjoy the creation of a finished product. The economic process, they asserted, should render direct satisfaction to the actual producer and not merely serve the wants or passing fancies of the consumers. Some present-day employers, who have become concerned about personnel problems of management, have tried to interest the workers in their tasks by giving them a feeling that they are parts of a cooperative whole and are sharing in a socially useful function rather than being mere drudges, concerned in selling so many hours of labor at so many cents an hour in order to be able to enjoy life in other ways.

List and the New Economic Nationalism

Friedrich List (1789-1846) followed the lead of Adam Müller with a much more systematic and sober body of ideas. Born in Württemburg, one of the small states of Germany, he began life in his father's tanning business but found it dull and went into public service. There he was able to devote time and means to acquiring a university education. In 1817 he was appointed professor of economics and political science at the University of Tübingen.

At that time the German states were united only by a loose federation, and each imposed a protective tariff of its own. List saw that each of these states was too small to prosper with a self-contained economy. He had become convinced that Europe under Napoleon had benefited from internal freedom of trade, with external protection against England's more highly developed industry. This, he thought, had provided a needed stimulus to German manufacture. He therefore advocated abolition of the import duties of the separate states and establishment of a customs union within Germany, protected by a moderate tariff for the nation as a whole. This unpopular stand lost him his university position in 1819. Three years later his advocacy of reforms brought a sentence of ten months' imprisonment for sedition. He was released before serving his full term on his promise to emigrate, and in 1825 he went with his family to the United States. List's idea of a German customs union was put into effect in 1833 under the leadership of Prussia and became a basis for the later unification of Germany under the Hohenzollern empire.

In America, List traveled about the country with Lafayette, settled on a Pennsylvania farm near Harrisburg, moved to Reading to edit a newspaper, discovered an anthracite coal mine, and built a railroad to connect it with the Schuylkill canal so that its product could be marketed. Pennsylvania was the heart of the agitation in behalf of protective duties in the United States, and List both spoke and wrote in their favor. He returned to Germany in 1832, put his doctrines in order, and published them in his major work, *The National System of Political Economy* (1841).

This work combined three important ideas, weaving them together into a cohesive system.

One was that the power of a community to produce wealth was not a mere matter of individual self-seeking but an organic, or, as many would say today, a cultural, situation favorable to production. "The prosperity of a nation is great, not in proportion to the accumulation of wealth, but in proportion to the development of productive forces." These productive forces include a desirable variety of natural resources, science and the arts, good laws, a high level of intelligence, the maintenance of order, morals, and a harmony and balance of the various industries and occupations themselves. To illustrate he cited as some of the great productive forces the Christian religion, monogamy, the discovery of the alphabet and the invention of printing, the postal system, improved transport, freedom of conscience, publicity of legal proceedings, parliamentary legislation. "It is difficult to conceive . . . of a law which can fail either to increase or to diminish the forces of production."

List's second basic contribution had to do with the striking inequality of capacity to produce among the world's great nations—a phenomenon as obvious when he wrote as it is today. From it he drew conclusions quite different from those of Adam Smith and his followers. The classical prescription was merely to let private enterprise spread; it would automatically do so as every nation concentrated on what it was best fitted to produce and sell. This doctrine was all very well for England, List thought, since England was first in the field and had already a well-developed and highly productive industrial community. But the younger nations, like the United States and Germany, would be prevented by British competition from developing a balanced productive base if they applied free trade.

For prosperity of coal mines, iron-smelting works near at hand were necessary; smelting works were of no use without rolling mills; rolling mills needed a market in the form of factories to make machines, and of railroad and building construction. Not one of these enterprises could make headway without each of the others. A more highly developed competitive nation could prevent the growth of the whole complex in a new region by underselling any one of its essential factors

The third important contribution by List was his use of the historical view of national and cultural growth. A policy that would be well adapted to one stage of development would not do for another. List admitted freely the beautiful logic of the classical free-trade doctrine, but he argued that it could fruitfully be applied only after all parts of the world had become well developed. World citizenship at the moment was impossible.

A nation in its early stages of industrialization needed to protect the industries required for a harmonious productive complex. At the cost of higher prices at the beginning for the products of protected industries, this nation could eventually enjoy lower prices after the industries had become established. The duties could then be removed. A highly developed nation could appropriately practice free trade; a highly developed world might, sometime in the far future, apply laissez faire. This reasoning laid a broad basis for the noted "infant-industry" argument for protection, advanced previously in the United States by Alexander Hamilton. The United States and Germany both applied the policy favored by List; he derived it from study and experience in these very nations.

Recent Economic Nationalism

List's doctrine, it will be observed, stressed absence of trade barriers *within* a national political unit as well as protection at its frontiers. What he had in mind was to unify and strengthen a state capable of nourishing an efficient productive community. The separate German principalities which he wished to unite were not large enough and did not include enough resources for this purpose. In the United States neither List nor those influenced by him would have favored import duties imposed by the separate states. He held that the absence of obstructions to commerce within so wide and richly endowed an area as that of the American Republic was an element of economic strength. And, if he lived today, he probably would now favor a low-tariff policy for the nation as a whole.

The arguments for protection which List used were disinterested in motive, but, like other bodies of economic doctrine, they were used by groups who would benefit from them, not only in Germany and the United States but in many other nations. Infant industries, grown to giants, did not want to abandon the favors bestowed by their nursemaid governments. Nations needing more power to maintain their existence at the beginning sometimes became, by virtue of exaggerated nationalism, international ogres in their later years, a did Germany itself. As in the case of the individual mar industries or nations which fail to outgrow the dependenc of childhood are likely to become predatory.

Observation of the outcome has induced a strong reactio among recent economic thinkers in favor of free trade an international laissez faire. Indeed, the world now is muc closer to the possibility of fruitful economic unity, if onl theoretical economics were concerned, than it was when Lis wrote. Yet the changed situation should not obscure the merit of his thinking or the fallacies he pointed out in classica reasoning.

Since World War II many have been concerned with th importance of extending the benefits of high productivity t the remaining underdeveloped regions of the world. Thos who study the problem almost invariably encounter the fac that the people of these regions want political autonomy an desire protective measures for infant industries, just as di the United States and Germany in the 1830s. Opinions diffe as to whether protective tariffs would be beneficial in suc cases, but the type of argument List used certainly has th same appeal in appropriate situations that it had in his day

Still more fundamental is List's stress on the necessity of general cultural basis favorable to high production. Again an again it has been discovered that a nation cannot much benefi merely by attracting large foreign loans and putting up highl mechanized factories, unless more elementary cultural an material advances have created the readiness to make goo use of these facilities. Education, health, desire for improve ment, reasonably honest and orderly government, a harmoni ous interrelationship among agriculture, manufacture, com merce, and good transportation, are all found to be necessar if the Industrial Revolution is to make much headway o yield much gain in any region. Most of these desired condi tions cannot be promoted without a large measure of activit by the national state. This is as true today as it was whe capitalism was replacing feudalism in Europe during the me cantile era.

What is new is that doctrines like those of the socialists an radical reformers are also widely known and command a legiance, so that when government intervenes it must, on per of being overthrown, intervene not merely in behalf of pr duction and profit, but also in behalf of fair distribution of th product and a reasonable degree of security for those wh do the work in the fields, the mines, the factories, and th railroads.

New regions now may conceivably be developed without
experiencing the centuries of slow growth, misery, and con-
t that characterized the birth and rise of industrial capital-
. By learning from history they may telescope the process
l pass by some of the blind alleys. But they cannot avoid
necessity of building the foundation and erecting the
ong frame if the house of modern civilization is to stand.

e Careys and American Optimism

Mathew Carey (1760-1839) was an early leader in Ameri-
economic thinking, strongly influenced by List. He be-
me a vigorous advocate of protection. He and his son Henry
793-1879) also added something of their own—an opti-
stic view as opposed to the pessimism of Malthus and Ri-
do. Industrialism, they strongly believed, not only could
ke nations rich but could benefit the common man. In this
y foreshadowed a strong current in American thought
ich has survived to the present.

Mathew Carey was born in Dublin, was dropped from his
rse's arms and lamed for the rest of his life. He developed
alent for languages and, much against the opposition of his
her, a butcher, decided to enter the printing business and
come a writer. He went to Paris, met Lafayette, and worked
Benjamin Franklin in his Paris printing shop. After learn-
his trade he returned to Ireland, started a journal critical
the government, and was imprisoned for sedition. On his
ease, in order to escape further prosecution, he smuggled
mself out of the country disguised as a woman and went to
iladelphia. Aided by Franklin's favor and an unexpected
t of four hundred dollars from Lafayette, he bought a
nting press, and in the course of time struggled upward
om poverty to become a leading publisher. Like many im-
grants, he cherished an exalted faith in his adopted country
d devoted himself to advancing the interests of the people,
pecially the Irish immigrants and the sweated clothing
rkers.

Mathew Carey was a charter member of the Philadelphia
ciety for the Promotion of National Industry. He developed
argument, not merely in favor of protection for manu-
cture, but on the basis of the same general concept as List's:
necessity of a well-balanced and well-integrated produc-
e system. In this system he stressed the importance of
riculture, for which he advocated protective tariffs, though
failed to see that American agriculture could scarcely bene-
from protection as long as it had a surplus for export. This

type of thinking lay behind Henry Clay's proposal for a "American System" in which the federal government wou subsidize and build internal improvements (roads and c nals), would protect industry, and would execute a liber policy of land distribution.

Henry C. Carey, though not distinguished as a technic theorist, had some revealing insights. He did not accept R cardo's theory of rent because, according to his observatio the first settlers often did not take up the best land but i ferior tracts. He foresaw, truly, great advances in agricultur technology and therefore in the productivity of land. He ther fore denied the validity of the "law" of diminishing return

Man's contribution to wealth, in Henry Carey's eyes, ove shadowed what nature held in store. And the riches obtaine from land were no more fundamental than those obtaine from industry; both depended mainly on the application (capital. The soil was merely one of the means of productio it would of itself produce little unless properly cultivate The return to landowners therefore was not distinguishab from the return to owners of capital or the payments for labo However shaky Carey's theory may be on this point, it corr sponded with the situation in the United States at the tim when land was abundant and the labor and capital necessa to make proper use of it were relatively scarce. Undevelope land was of little value to anybody.

Henry Carey also rejected, and on similar grounds, t Malthusian theory. The technical capacity to produce wou increase, not more slowly than population, but more rapid In his *Principles of Social Science* (1858-1859), he observe shrewdly that "fecundity is in the inverse ratio of organiz tion" and that the "cerebral and generative powers of m mature together." In other words, the birth rate falls in i dustrial societies, while man's ingenuity is capable of increa ing the means of subsistence more rapidly than the popu tion. This observation has been proved true in the Unit States, even long after the passage of the frontier and t end of the expansion of acreage in farming.

The optimism of Henry Carey's view extended even distribution of the increasing product. Interest and rent wou fall as the years passed, he believed, not as a total amou but as a percentage share of the rapidly increasing produ Hence the workers would gain greatly from the increase per-capita output. The gain of labor, however, would not at the cost of robbing capitalists of their necessary incenti to increase production. Rather the contrary; there would ex a harmony of interests. No theme is more characteristic recent American economic publicity than the effort to convince both capital and labor of their harmony of interest in rapid increase of productivity and widespread distribution of the resulting product. Though both often act as if this theme were untrue, few fail to give it lip-service.

Henry Carey was not a reckless optimist; he should be honored as one of the first to advocate conservation of natural resources, especially the soil. All mineral constituents taken from it in raising crops must be restored, he pointed out, if it is not to be ruined. "It is singular," he wrote, "that modern political economy should have so entirely overlooked the fact that man is a mere borrower from the earth, and that when he does not pay his debts, she acts as do all other creditors, and expels him from his holding."

The Careys did not include the commercial or trading groups in their harmony of interests; they wanted to diminish the toll of middlemen by decentralizing industry, so that farms and factories would be near together. In this they reflected a prevailing agricultural prejudice from that day to this. In other respects, they favored laissez faire. As social thinkers, however, they traveled a long way from the atomistic individualism of the classical economists.

Henry George and the Single Tax

By far the most famous American economic writer, author of a book which probably had a larger world-wide circulation than any other work on economics ever written, was Henry George, author of *Progress and Poverty* (1879). This book expounded a theory developed with superb logic. Its author followed the rules for writing which he later gave his son: make short sentences, use small words, avoid adjectives, shun "fine writing." He did not include in these instructions all that made his own work read, however—eloquence springing from deep conviction and a talent for explanation by use of apt illustrations and references to common experience. Above all, it was an attack on a grave problem, which worried millions at the time of its publication in the depressed 1870s, and it recommended a concrete remedy.

Henry George's ultimate influence on economic theory and public policy was, however, small in comparison with the sensation he caused during his lifetime. The quality of his thinking has probably been underrated by professional economists. This is curious, in view of the fact that his method is precisely the same as that of the great classicists like Adam Smith—deductive reasoning bulwarked by shrewd observation. The remedy he proposed was, moreover, based squarely on the most prominent conclusion of Ricardo—that rent was unearned. Indeed, he succeeded in improving the Ricardian theory of rent.

Born in 1839 in Philadelphia, Henry George was much influenced by his mother, a former schoolteacher who was religious and sensitive. He did not do well in school and at the age of sixteen shipped before the mast on a sailing vessel. During his two years at sea he saw something of Asia. After his return he started to learn printing, had trouble with the management, and was discharged. Then he set out for California, where by 1856 the gold discovered in 1849 was already running thin. It was hard to find work in printing shops, still harder to find gold on the occasions when he was without a job. Eventually he got a start on a San Francisco newspaper and rose rapidly from printer to editor. All this time he had been a habitual reader—particularly of poetry—and had spent long hours cultivating his prose style. One article was accepted by the *New York Tribune* and praised by John Stuart Mill. He helped to found the *San Francisco Post* in 1867; it collapsed in the panic of 1873.

George was deeply impressed by the struggles with poverty which he had experienced and which he saw all about him in so rich and rapidly developing a country. He carefully read the celebrated economists, but was still puzzled by "the persistence of poverty amid advancing wealth." He observed acutely the conditions in California. One day, he writes, during a conversation the answer came to him in a flash. He devoted the rest of his life to elaborating, refining, and promulgating it. The book in which he gave full expression to his doctrine should have made his fortune, but it was rejected by publishers. Set up by a friendly printer, and by Henry George himself, it was finally sold outright for a small sum to the printing house which issued it. It did, however, bring him fame. He moved to New York, ran for mayor in 1887 on a labor ticket, coming in second to the Democrat and defeating the Republican Theodore Roosevelt. In 1897 he ran a second time, but died before the election.

Henry George's Theory

George observed that the worst pauperism, as well as the greatest riches, could be found in the oldest centers of industrialism—that is, where the most progress in power to multiply man's production by modern techniques had occurred. Newer regions suffered less from this difficulty, though they were not so rich. This was just the opposite of what one would expect. Why should the mass of humanity not benefit from the great gain in industrial productivity? The answer must lie in the distribution of the wealth produced.

He examined the generally accepted doctrines of distribution—the theories which were supposed to explain the division of the product among the three chief factors of production: labor, capital, and land. He found the accepted theories of wages, profits or interest, and rent all unsatisfactory and without much relation to one another.

The classical theory of wages as explained by Adam Smith and developed by most of his followers declared that the maximum possible payment to wage-earners was determined by the fund of capital devoted to productive enterprises. This "wages-fund" theory held that the capitalist advanced money to the wage-earners so that they could live while producing. The capitalist then recouped his investment in labor by selling the product. George concluded that this was putting the cart before the horse. "Wages, instead of being drawn from capital, are in reality drawn from the product of the labor for which they are paid."

Capital, he asserted, does not have to be accumulated prior to production by labor in a "natural" society. "If, for instance, I devote my labor to gathering birds' eggs or picking wild berries, the eggs or berries I thus get are my wages." The situation in a modern society where division of labor prevails is not essentially different. Through a complicated mechanism of exchange, the workers support one another while all are working. Capital alone could not feed or clothe any workers if they should all stop working. Indeed, the value of capital itself would in that case disappear. What capital really does, and what it gets paid for, is to supply the improved machinery and other facilities which enable the workers to be more highly productive.

The Malthusian theory was supposed to prove that wages would remain near the minimum of subsistence, since population growth would always tend to outstrip the food supply. But the theory, said George, was not in accord with what we know about population or with what we know about production. Total wealth has increased faster than population in all industrial nations. "The richest countries are not those where nature is most prolific, but where labor is most efficient—not Mexico, but Massachusetts; not Brazil, but England." "Both the jayhawk and the man eat chickens, but the more jayhawks the fewer chickens, while the more men the more chickens." A large population in a given area produces more efficiently than a small one because it is susceptible to greater division of labor and can make use of more advanced techniques.

The poverty of labor is therefore the result of unjust distribution, not of any natural law governing wages or population. Is it, then, due to unduly high returns to capital? Not if the returns to capital are properly analyzed. Strictly speaking, the return to capital might be regarded as the rate of interest. But interest rates are highest when wages are highest, not the reverse, as one would expect if the investor of capital was gaining by exploiting labor. It is precisely in the most highly developed countries, where disparities of wealth are greatest, that interest rates are lowest.

And what of profits? Well, profits is a word that covers a number of different kinds of return. Part of profits is pay for the work of the employer: in other words it is, strictly speaking, payment for labor. Part is, like interest, payment for use of capital, but the rate of return represented by this part must be about the same as interest, though perhaps higher than the going interest rate because of the risk involved. The remainder is payment for use of land or other natural resources, in other words, rent.

By a process of elimination, George thus arrived at the payment of rent as the cause of poverty. In Ricardo he found at least a partial explanation—only the better lands command rent, and this payment is for nothing except possession. Having discarded the Ricardian theory of wages and profits, however, George disagreed with Ricardo that landowners exploit only the capitalist; what they really do is prevent labor from benefiting to the full by modern methods of production. Even this conclusion he qualified: workers can benefit somewhat if the rate of technical progress keeps ahead of the rise in monopoly value of good land. This is the case in newer regions, he believed, but not in older ones.

At this point in his reasoning George made an original contribution which has stood the test of time. What really brings the most increase in value to land is not its differential in fertility but the growth of population in the neighborhood and the general increase in the productivity of society. To get rich, buy not the best farm land, but a tract which will turn out to be near the center of a growing city. It makes no difference whether this tract is covered with the most fertile loam or is solid granite. Who knows anything about the agricultural value of the corner lot at Broad and Walls Streets, New York?

The Remedy Recommended

Since rent properly defined—that is, the payment for use of land or natural resources themselves, not for buildings or other improvements—was unearned, George thought that land ought to be public property. Yet he hesitated to advise so drastic and revolutionary a procedure as confiscation. The same purpose could be achieved, he thought, by a tax which would appropriate for public use all rise in the value of land, or the "unearned increment." The unearned increment would, he thought, be so great as to pay all governmental expenses; therefore all other taxes could be repealed. This would encourage trade and industry and lift from the workers the heavy burden of taxes on production and consumption. Also, the fact that nobody could profit merely by holding land would force landowners to improve their property and so would increase the total of production and stimulate competition. Elaborating on the virtues of the "single tax" at great length, George was able to predict a virtual paradise if it were only to be adopted. It would mean full and steady employment, the abolition of slums, the steady rise of wages through more rapidly expanding demand for labor.

Why was not so appealing and logical a proposal ever put into effect? Economic theorists promptly began to pick flaws in it, but they were not nearly so widely read or eloquent as George. What he had done was, in effect, to issue a challenge to the going economic system almost as fundamental as that of Marx. Marx wanted to abolish private ownership of capital; George wanted to abolish at least the gains from private ownership of land.

But landowners were as powerful as owners of industrial capital. In the United States they were even more powerful, since ownership of land was widely distributed. Speculation in land values had been, in this new country, the standard way of accumulating wealth since the beginning; the attempt to gain in this way had been practiced not only by those already possessed of wealth or those who made real estate their business, but by almost every settler in a new region and almost every farmer. Even urban labor had been prominent in the agitation for liberal distribution of publicly owned land. George was opposing not merely a limited group of exploiters but the prevailing folkways of a whole people.

A few local governments did try limited applications of the plan, but it never made much headway. In fact, the only serious and general use of the scheme was inaugurated by a socialist government in Britain after World War II. Even here it dealt only with urban land or land near towns, and the complications of devising a fair application were enormous. Moreover, the British socialists did not rely on it to solve all economic problems. They were conscious that much more than this had to be done to approximate their goal of social justice. The concept of the unearned increment, however, has found a secure place in social thinking.

Was Henry George Right?

The case of Henry George constitutes an illuminating example of the pitfalls that threaten even the most observant, logical, conscientious, and humane thinker in the social sciences, so long as he is limited to rough observation and deduction from a few simple premises. Such theories need to be tested by more comprehensive and accurate assessment of the actual situation than it was possible to make when he wrote.

Statistical studies of the national income in the United States, of a sort which had never been made and scarcely were conceived during his lifetime, reveal with a high degree of certainty that, at least in the period between the Civil War and the present, the real income of labor in the United States *has* kept pace with the advance in power to produce. It has not done so with absolute regularity, but it has approximated the same gain over the years. The same body of information indicates that the share of property income has *not*, over any considerable period, increased at the expense of labor income. A complete abolition of "unearned increment" could not have made an important difference.

The very problem which Henry George set out to solve therefore did not exist, in the form in which he stated it. What he was really observing was an unusually prolonged depression succeeding a period of rapid industrial expansion and heavy immigration. The unemployment and the misery were there all right, as well as the extreme inequality in distribution of income. But what he ought to have set out to analyze was the cyclical instability of the economy, to which the classical economists had paid little attention, rather than the classical "laws" of distribution.

Logic is an essential part of scientific method, and so is a body of theory. Yet the best logic and the most highly consistent theory may be irrelevant to the behavior of what we call "nature." To choose really strategic problems, and to apply to them analysis that can be carefully tested by relevant methods of observation, is of supreme importance if we wish to develop a reliable and useful science.

6

The Main Current—
Classicism Reworked

The impression the classical economists had made on think-ing, particularly in England, France, and the United States, was too strong to be erased by the attacks of socialists either utopian or "scientific," or by the varied doctrines of those who assaulted the orthodox methods. It persisted in university teaching and in the "respectable" circles of society. The eco-nomic protestants were regarded as mavericks, men doubtless to be honored for their humanitarian impulses but belonging to a sort of underworld of science.

Scholars went on trying to build on the foundations laid by Adam Smith, Malthus, and Ricardo, refining and remodeling, adding this wing or that to the structure, but not scraping the main plan. The strength of the tradition was buttressed by the fact that capitalist industrialism, for which the classical doctrine provided a rationale supposed to be derived from laws of nature, was still young, vigorous, and rapidly growing. Though subject to bitter antagonism, capitalism did not col-lapse, nor was it essentially changed by attempts to introduce small-scale utopias that might successfully compete with it.

Manchester School and Tariff Repeal

The classical doctrine of laissez faire in international trade never was generally adopted and was not approximated in practice even by the nation where it was most ardently preached, Britain, until years after Smith and Ricardo had espoused it.

England protected the home-growing of grain by the Corn Laws. (The English mean by corn not Indian corn, as do the Americans, but grain.) In 1815, at the close of the Napoleonic Wars, no wheat could be imported below the high price of ten shillings a bushel. Though this wall was later somewhat lowered, tariff reformers were not able to reduce it nearly as much as the protective duties on manufactured products and other raw materials.

The hereditary ruling classes were large landowners, and their chief income came from agriculture. Newly acquired wealth, of which there was plenty as a result of war inflation, was largely invested in land. The English manufacturers, who were getting rich too, did not need or want protection since, being first in the field, they had little competition from abroad and enjoyed a wide export market. Yet the power of the landed aristocracy was still dominant, buttressed as it was by a limited franchise.

Every time a poor harvest occurred the price of food climbed and the wage-earners suffered near-starvation. If the manufacturers raised wages to make up for the high price of food, their profits suffered. Both employers and workers therefore opposed the Corn Laws, but they could make little headway until after the passage of the Reform Bill in 1832 extended the right to vote. Poor crop years in 1836 and thereafter, accompanying world-wide depression, stirred the old grievance deeply.

An association was formed in Manchester to combat the Corn Laws. Richard Cobden, a self-made cotton manufacturer and merchant, who had accumulated enough money to retire and study, threw himself into the debate with all his talent and vigor. Later he succeeded in enlisting the active assistance of John Bright, who became a noted orator and statesman. They, with others, carried on a great public campaign, financed and organized a "grass roots" political machinery which captured the House of Commons. No more convinced and effective advocacy of free trade has ever been known. Robert Peel, Prime Minister, whose father had been an industrialist as well as a landowner, was sympathetic. In 1845 the cause was helped by a failure of the English wheat crop, followed by the potato famine in Ireland. The next year the landlords' opposition was at length overcome, the Corn Laws were abolished by 1849, and Britain adopted a general free-trade policy that endured for more than seventy years. Thus history seemed to vindicate the classical economists. Industry had at last conquered feudal agriculture in the United Kingdom.

The prosperity which on the whole characterized Britain

uring the period of her great industrial expansion in the
iiddle and later years of the nineteenth century made possi-
le a rise in real wages and the extension of liberal democracy.
ocial conflicts were by no means at an end, but moderation
nd reason thrived in the new climate. Cobden and Bright had
dded little to economic theory, but their eloquence, com-
ined with the good fortune that for so long seemed to attend
eir victory, brought them fame as founders of the Manches-
r School, pre-eminent followers of the classical free-trade
octrines.

ohn Stuart Mill's Revision

It was in 1845—the year before Cobden and Bright won
eir victory in Parliament—that John Stuart Mill began to
rite his *Principles of Political Economy.* The book was pub-
shed in 1848, the same year the *Communist Manifesto* by
1arx and Engels appeared. Mill, like Marx, had been deeply
ifluenced by the writings of the utopian socialists and other
ritics of capitalism, and he had observed the turmoil of the
mes, but, unlike Marx, he foresaw no necessary debacle for
ie capitalist order as a whole. Nevertheless he believed it
ecessary to restate economic theory, including such additions
nd changes as would make it not a consistent whole but
ould allow room for indefinite human betterment.

No economist ever had a more generous spirit than Mill,
one showed a finer sense of order, none was master of a more
icid prose. His reformulation of the classical doctrine was
broad and tolerant one, which called for calculated inter-
ention to spread the benefits of progress. The book was im-
iensely successful and dominated the field for years. Mill
as now read instead of Smith and Ricardo.

John Stuart Mill (1806-1873) was the oldest son of James
Iill, the economist. The father evidently decided that his
oy was to be unexcelled in scholarship. John began to study
reek at the age of three and had read widely in the language
y the time he was eight. Then he went on to Latin, at the
ime time teaching his brothers and sisters, learning mathe-
iatics, and giving his father oral reports on the histories and
eatises on science which he read. These reports, rendered
uring a daily walk in the early morning, began to include
dam Smith and Ricardo when young Mill was thirteen.
'ather and son would discuss the contents of the books, clari-
ying the obscure points. James Mill's ideas for his own book,
olitical Economy, were elaborated by discussion in the same
ay, before the final text was written and published.

After a year in France, John returned and studied law. In 1823, when he was seventeen, he obtained a minor position in the East India Company. He rose rapidly and soon achieved a post that gave him both income and leisure to carry on independent work. When the company lost its charter in 1858 he retired on a pension of £1500.

At the age of twenty-two the younger Mill met Mrs. John Taylor (Harriet Hardy), then twenty, with whom he formed a close intellectual companionship and who greatly influenced his thinking. Twenty-one years later they married. In many of Mill's works the reader will find passages that reveal his admiration for women and his respect for their accomplishments. Mrs. Taylor did much to inspire his essay *On Liberty* and his later feminist tract *The Subjection of Women*.

Aside from works on economics and public affairs, Mill wrote on logic. In his formative years he moved in a cultivated circle which included Ricardo, Jeremy Bentham, Carlyle, and Macaulay. An intellectual in a relatively calm and secure environment, apparently happy in his life and work, Mill exemplifies the temper of nineteenth-century British liberalism at its best.

Mill's Basic Doctrine

The major change that Mill's *Principles* introduced into the classical tradition was the assertion that the distribution of wealth was not governed by unalterable natural law but could be influenced by the will of man.

The classical ideas about production, somewhat modified and broadened, were, he held, universally valid and could be applied to any form of society. But the distribution of wealth "is a matter of human institution solely." It "depends on the laws and customs of society. The rules by which it is determined are what the opinions and feelings of the ruling portion of the community make them, and are very different in different ages and countries; and might be still more different, if mankind so chose."

Probably there are natural laws which govern the formation of opinions as well, Mill thought, but he regarded this realm of inquiry as far more difficult than the narrower one of political economy and did not venture to speculate about it. What the economist could do was to trace the way in which any given system was likely to work. "Society can subject the distribution of wealth to whatever rules it thinks best; but what practical results will flow from the operation of those rules must be discovered, like any other physical or mental truths, by observation and reasoning."

At one stroke Mill therefore removed the hopeless aridity of classical doctrine, which seemed to condemn men to observe a supposed natural law inhering in private enterprise under capitalism—a law that had been held to perpetuate poverty and injustice.

The Laws of Production

Mill's exposition of the economic laws of production added little to Adam Smith in essence, though much in range. The basic factors of production were, he said, labor and natural resources. Both concepts, however, he much broadened. Labor was both bodily and mental; it included not only the contribution of the manual worker but that of the inventor, the philosopher, and the statesman, and not only physical effort but its emotional setting. Natural resources included not only materials that might be fashioned but power to aid men in fashioning them, or even to be a substitute for labor. "The powers of nature," he declared, "do all the work when once objects are put into the right position. This one operation, of putting things into fit places for being acted upon by their own internal forces, and by those residing in other objects, is all that a man does, or can do, with matter." Men had acquired a command "over natural forces immeasurably more powerful than themselves; a command which, great as it is already, is without doubt destined to become indefinitely greater." A prophetic statement indeed.

With these views, Mill naturally did not overlook the productive function of what we now call "pure" science. "The labor of the savant, or speculative thinker, is as much a part of production in the very narrowest sense as that of the inventor of a practical art; many such inventions having been the direct consequences of theoretic discoveries." Mill also assigned a productive function to consumption. A good diet helps the producer, a good education the thinker. There may be unproductive consumption too. This is not to be disapproved if it adds to the joy of life; it may be regarded as a true economic surplus.

Capital Mill sharply distinguished from money. Capital is an accumulation of goods that facilitate further production. What capital does for production is to afford the shelter, protection, tools, and materials which the work requires, and to feed and otherwise maintain the laborers during the process." It is the product of past labor. Capital in this real sense does not pile up indefinitely; it has to be continually renewed and replaced, though not so rapidly as goods for immediate consumption. "The capital needs not necessarily be furnished by a person called a capitalist."

Mill on Distribution

Private property, Mill held, was an institution introduced into primitive communities not because it was useful to society, but merely to maintain peace when quarrels broke out. To suppress aggression, the primitive tribunals "gave legal effect to first occupancy, by treating as the aggressor the person who just commenced violence, by turning, or attempting to turn, another out of possession." What the possessor had when trouble broke out was not necessarily the product of his own labor.

But private property was not the only conceivable basis for distribution of wealth. Mill reviewed the programs of communists and socialists, the plans for associationist or cooperative colonies. Whatever their merits or defects, such schemes "cannot truly be said to be impracticable." Individuals, to be sure, might try to evade their fair share of the work. "The objection supposes that honest and efficient labor is only to be had from those who are themselves individually to reap the benefit of their own exertions. But how small a part of all the labor performed in England, from the lowest paid to the highest, is done by persons working for their own benefit."

A greater difficulty would be "to apportion the labor of the community among its members." But even this difficulty might be overcome. If the only alternative to communism were "the present state of society with all its sufferings and injustices," then "all the difficulties, great or small, of communism, would be as dust in the balance." It is scarcely necessary to point out that Mill was not writing of the doctrine now called communism or the present regime in the Soviet Union.

Mill went on to describe how wealth is distributed under capitalism. Though he used much the same concepts as those of the classical theory of distribution, such as the wages fund and the Malthusian population theory, he pointed out possibilities for advance. The real minimum which workers will accept, unless driven by desperation, is not a bare subsistence but the standard of living to which they are accustomed. Though Ricardo was right, he thought, that profits must fall if costs of labor rise, costs of labor are not identical with wages. "The cost of labor is frequently at its highest where wages are lowest. . . . Labor, though cheap, may be inefficient." Labor cost drops when workers become more efficient; workers may also be more efficiently used because of technical advance. Population growth may be restrained. Large profits and high wages may exist together. This actually has been the tendency in Britain and the United States since Mill wrote.

Mill, however, made virtually no advance beyond the Ricardian theory of rent or the dry and involved classical ideas about value and price. But he was careful to note that the doctrine of price was peculiar to the type of distribution prevailing under capitalism and, rather than being a "law of nature," might be altered. And, he warned, "We must never forget that the truths of political economy are truths only in the rough."

The Progress of Society

The most interesting part of Mill's *Principles* is its fourth and last part, "Influence of the Progress of Society on Production and Distribution." This conclusion begins by denying the desirability of mere material progress as an ultimate goal. More and more people vying with one another for more and more wealth do not present a pleasing spectacle to the author, even though they might progressively achieve their ambitions. As a horrible example Mill points to the free, democratic, and relatively prosperous "northern and middle states of America," where "the life of the whole of one sex is devoted to dollar-hunting, and of the other to breeding dollar-hunters." He admits that people should not rust and stagnate, and that "while minds are coarse they require coarse stimuli." But we are in a "very early stage of human improvement." He foresees an ultimate "stationary" state of society, in which population growth will be voluntarily limited, there will be ample means to maintain all in reasonable comfort, and attention will be directed to a better distribution of the product rather than merely to increasing it without limit.

Why fill the world with as much population as it will hold, even if, because of technical progress, this population might be adequately fed? "It is not good for a man to be kept perforce at all times in the presence of his species." There is no satisfaction in thinking of "every rood of land brought into cultivation which is capable of growing food for human beings; every flowery waste or natural pasture plowed up, all quadrupeds or birds which are not for man's use exterminated

as his rivals for food; every hedgerow or superfluous tre
rooted out, and scarcely a place left where a wild shrub o
flower could grow without being eradicated as a weed in th
name of improved agriculture."

Rather, increased power over nature should be used t
shorten hours of work, and the time gained should be devote
to cultivation of other things. There would be much mor
likelihood of improving the Art of Living "when minds cease
to be engrossed by the art of getting on."

As for labor, it should not be patronized, and should not b
regarded as an inferior class needing to be patronized. "
do not recognize as either just or salutary a state of society i
which there is any 'class' which is not laboring." But "th
working classes have taken their interests into their ow
hands. . . . To their own qualities must be commended th
care of their destiny. . . . The prospect of the future depend
on the degree in which they can be made rational beings.
Mill correctly foresaw, and approved in principle, the strengt
of the independent labor movement, acting both industriall
and politically to improve the condition of the workers an
bring about the economic goal on which he laid major stres
—a more nearly equal distribution of income.

Nor did Mill expect that capitalism as he knew it woul
endure forever. "It is not to be expected that the division o
the human race into two hereditary classes, employers an
employed, can be permanently maintained. The situation i
nearly as unsatisfactory to the payer of wages as to the re
ceiver." The economic process is essentially a cooperative on
and it should be placed on a cooperative basis. Mill was hop
ful for the future both of producers' cooperation and co
sumers' cooperation. The first has not fulfilled his hope; th
second has grown into a powerful, world-wide institution.

Government could be used, he held, to promote the proces
of change which he desired. He favored inheritance taxes fo
the purpose of equalizing income. Government already di
much more than act as a policeman: it took an interest i
conservation of resources and should take more, since natur
resources are a common heritage of the human race; it d
cided civil disputes not involving violation of statutes; it coine
money; it undertook public works. Mill thought the doctri
of nonintervention by government fallacious; extensions o
governmental activity should be judged by its expediency i
any given case.

as Mill a Socialist?

In his *Principles,* Mill stated that he sympathized with the cialists in their practical aims and thought the time ripe for troduction of these changes, but he disagreed with their tack on competition. "They forget that wherever competition not, monopoly is; and that monopoly, in all its forms, is the xation of the industrious for the support of indolence, if not plunder." Competition was even good for laborers when mand for labor exceeded the supply. Though cooperation uld be substituted for it in associations, these associations uld and should compete with one another. "It is the common ror of socialists to overlook the natural indolence of man- nd; their tendency to be passive, to be slaves of habit, to per- t indefinitely in a course once chosen." Mill did not wish mpetition to prevent people from earning the means of a cent livelihood, but he emphasized the necessity of a com- titive spur. What he did disapprove was advancement of a ivileged group at the expense of the majority.

In his *Autobiography,* published after his death, Mill wrote himself and his wife, "Our ideal of ultimate improvement nt far beyond Democracy, and would class us decidedly der the general designation of Socialists. The social problem the future we considered to be, how to unite the greatest dividual liberty of action with a common ownership of the w material of the globe, and an equal participation of all in e benefits of combined labor." A difficult problem indeed, ward a solution of which the "mixed" economies that more d more characterize the advanced industrial states seem to feeling their way. Mill was obviously neither a utopian nor Marxist.

Preceding bodies of doctrine had been adapted to defense , or aggression by, agriculturists or capitalists or revolution- y workers or utopian dreamers or proponents of national onomies. What sort of emergent force could use his leading eas? Clearly a socialistically inclined but pragmatic labor ovement, strongly inspired by cultivated intellectuals whose n moral and spiritual comfort depended on a society in ich they could feel more at home. Nothing has suited this nd of love of liberty, gradualism, economic egalitarianism, d aims beyond material advance better than the tradition of itish democratic socialism.

The line of classical legitimacy to which Mill clearly be- nged, and which in him reached its climax, thus has turned out to be a herald of the strange amalgam which society in Britain, where the Industrial Revolution first flourished, is by way of becoming. Mill, in his *Principles of Political Economy* published simultaneously with Marx's *Communist Manifesto*, seems to have been the truer prophet, at least for his own nation.

Classicism on the Defensive—Cairnes

After John Stuart Mill there were few eminent economists in the classical tradition who ventured to write in a broad philosophical vein or present in a single work a complete theoretical system covering all aspects of the "science." In that sense the doctrine entered a long decline. It fell mainly into the hands of academic specialists who narrowed its scope, busied themselves with technical elaboration of sections of the doctrine which they thought had been unsatisfactorily developed, or refurbished familiar ideas in a way that might better be defended from the criticisms aimed by many against the classical abstractions.

Among these one of the first was John Elliott Cairnes (1823-1875), born in Ireland, who was graduated from Trinity College, Dublin, and subsequently taught at Dublin, at Queen's College, Galway, and at University College, London. Cairnes again limited the proper subject matter of economics to production and distribution of wealth, excluding ideas of social policy. He also erected an elaborate defense of the deductive method of thinking.

Induction from facts of the real world is misleading, Cairnes held, because in the real world so many influences are at work simultaneously that it is difficult to tell which of them, or what combination, is responsible for the consequence in which the investigator is interested. Physical scientists can test the validity of inductive hypotheses by laboratory experiment, but experiment is impossible in economics because the world is not a laboratory.

The economist, however, can reason fruitfully by assuming certain oversimplified postulates, abstracted from reality, and then discovering by deductive logic the necessary consequences. For example, one may assume that "perfect competition" exists, that each businessman is trying merely to maximize gain in the sale of a single article, that he has knowledge of his own best interests, etc. etc., and on that basis one can deduce a "law" of price as related to demand and supply—as many subsequent economists have done. Such a process is, Cairnes thought, like "an experiment carried on mentally."

This type of device has saved for economists in the classical tradition their chief stock-in-trade. It has made possible more and more elaborate hypotheses. The charm of logical problem-solving is so seductive to many minds that they are almost unconsciously tempted to regard hypotheses so discovered as essentially true representations of what occurs in the world of practice. Careful thinkers using this technique will admit that modifications must be made in application of their conclusions, but they often overlook the possibility that their postulates are so far from being controlling influences in reality that their hypothetical abstraction has little or no relation to what actually takes place. They are continually tempted to substitute a never-never-land of their thinking for the world around them.

Cairnes used the device, in the main, to buttress classical doctrines like the wages-fund theory, then under telling assault. He thus fell into the error of supposing that trade unions can make no permanent gains in wages that would not have occurred had they not existed.

Cairnes did, however, introduce the pregnant idea into classical reasoning that labor is not always mobile or capital fluid, and therefore that competition works well only within limited groups.

The Resort to Mathematics

Once embarked on processes of deduction from postulates, it was inevitable that economists should resort to mathematics of the symbolic type—which is a neat way of utilizing logic dealing with abstractions. Equations and graphs can be used to express assumed relationships among economic factors without knowing any of the actual quantities at all.

For example, suppose one wanted to express the idea that a fall in the price of any given article will cause demand for it to rise, and that the supply of the article will be increased by rising prices. And suppose one wants at the same time to express the conclusion that the price of the article will tend to settle at a point where the demand exactly equals the supply (since a larger demand would raise the price and a smaller one would reduce it), while variations in the supply would have exactly the opposite effect.

All this can, of course, be said in words, but it can also be said by a graph, in which the vertical axis represents prices extending from zero as high as you like, and the horizontal axis represents variations in quantities. Then a curve D-D′ sloping downward to the right will illustrate demand increasing as prices fall, and a curve S-S' sloping upward to the righ will represent supply increasing as prices rise. Thus—

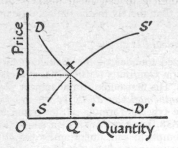

In such a graph, demand and supply are in balance wher the curves cross at x. The equilibrium price is P, and th quantity that will be produced and sold at that price is Q. Not that this purports to be a general statement of price-quantit relationships under the influence of demand and supply, with out the use of any numbers at all. It is a relatively simple dia gram; equations and diagrams used in economic reasoning ca be developed to a high degree of complexity. But equations an graphs do not prove anything about the validity of the assump tions on which they are based, or tell what would happen i any given case even if the assumptions are true, unless th actual numbers can be supplied.

Obviously one cannot get out of the mathematical machir anything not implied by what is fed into its hopper. It is mere a convenient way to do logical thinking and a shorthan method of illustrating the thinking process that might occur many more pages and be less accurately visualized if stated i words. But it is not a magic method of discovering truth.

Those economists who first began to employ mathemati are frequently called the "mathematical school." This is scarc ly an appropriate name, since they applied themselves to mar problems, beginning with the abstruse theory of value an prices, ranging through the theory of money, of productio and even distribution of wealth. These economists came widely different conclusions. Few economic specialists today without mathematics altogether, no matter what the type their thinking or the nature of their doctrines. But most those who have made extensive use of symbolic mathemati are, in the nature of the case, reasoning deductively from a sumed premises, just as did Smith, Ricardo, Malthus, ar

eir followers. They are also thinking largely as if social effects
re caused by individual behavior of "economic men." It is
erefore appropriate to call them thinkers in the classical
dition rather than merely mathematical economists.

rly Mathematical Classicists

The recognized founder of the mathematical method was
ntoine Augustin Cournot (1801-1877), professor of mathe-
tics at Lyon. His first book on the subject, *Recherches sur
principes mathématiques de la théorie des richesses*, used
ch the same type of formula as illustrated above. It was
blished in 1838, but for years not a single copy was sold.
bsequent attempts to gain attention by simplifying the ma-
ial were also unsuccessful. Cournot's economic writings
uld have remained almost unknown if they had not been
g up and publicized by the celebrated English economist
illiam Stanley Jevons. They contributed little to the sub-
nce of classical theory.

The same fate awaited Herman Heinrich Gossen (1810-
58), a German public official who thought his mathematical
tement of economic theories, *Die Entwickelung der Ge-
ze des menschlichen Verkehrs*, would bring him as much
ne as Copernicus had won by his discoveries of laws govern-
; the motion of heavenly bodies. Disgusted by the fact that
body would buy his book, published in 1854, Gossen de-
oyed the entire edition, but one copy had somehow reached
gland, was discovered by a professor in the British Museum,
d was, like Cournot's, brought to public attention by Jevons.

rginal Utility

Gossen did offer a contribution to theory which, independ-
ly duplicated by others, has been the delight of economic
thematicians ever since. He was the first to make a wide
plication of the marginal concept used by Ricardo in his
ory of rent. Gossen applied the idea to demand in general.
e outcome is a complex theory of price and value resting
Jeremy Bentham's pleasure-pain principle, and hence usu-
y called the theory of "marginal utility."

The major assumption underlying this theory is that the
ue of anything to the purchaser declines, the more of it he
. A hungry man would pay handsomely for one loaf of
ad—probably much more than it costs. A second loaf
uld give him less satisfaction; therefore it would be worth
s to him. Presumably he would go on buying loaves of

bread until he reached a point where the cost of paying th
price demanded caused him more pain than the pleasure h
would receive from eating the loaf. The last loaf he buys—th
marginal loaf—therefore determines the extent of his deman
But since he probably buys his day's supply of bread all at onc
and does not actually buy the loaves one at a time at decreasin
prices, the price of the marginal loaf determines what th
whole lot has to be sold for.

This man of course has many other wants besides that fc
bread. He applies the same calculation to everything he buy
Nobody, however, can buy everything he might want. Ther
fore the consumer apportions money for his various purchase
according to the relative marginal utility to him of each kin
of article.

It is easy to see that this type of theorizing can be applie
to almost any economic relationship involving purchase an
sale—to production, for instance, by thinking of the last un
which the producer is willing to make at the price he can g
for it; to wages, by thinking of the last labor-hour the employ
is willing to engage; to interest on capital, by thinking of th
last dollar the capitalist is willing to lend. Each party to eac
bargain weighs his satisfaction from what he can get again
the sacrifice of what he has to give in exchange. The totali
of all these influences in any market is supposed to underl
supply, demand, and price for the whole market.

The almost endless intricacies of such reasoning are heave
sent material for mathematical expression, and those who d
veloped it did not fail to make full use of the symbolic la
guage. Prominent among the pioneers were William Stanle
Jevons (1835-1882), an English professional economist, Léc
Walras (1834-1910), a French-Swiss teacher of economic
and the Austrians Karl Menger (1840-1921) and Eugen vc
Böhm-Bawerk (1851-1914). Jevons never finished his maj
work; he is known not only for his marginalist theory but fc
his probably erroneous suggestion that periodic crises may k
due to sunspots, acting through the weather on crop yield
Walras was a reformer who advocated nationalization of la
before Henry George did, but he is principally celebrated fc
the fact that he was the first to develop a complete mathema
cal system for a whole economy, showing how it would, und
his assumptions, tend toward equilibrium. His major assum
tions were a regime of private property and perfect compe
tion. Böhm-Bawerk, who was concerned with refuting Ka
Marx, is known chiefly for his theory of interest based on th
marginal utility of capital—a theory that introduced compa
son of present and future values.

The marginal theory was applied not only to markets and production but to distribution of income, notably by an American economist, John Bates Clark (1847-1938). The executive of a business, according to the analysis, combined land or natural resources, capital and labor, to produce a marketable product. How much would he pay for each of these factors of production? It would depend on their *relative* marginal productivity. The last unit of each which he believed it desirable to pay for determined its price. If a farmer found it worth while to employ more labor on a given piece of land, he would keep on hiring until the marginal labor unit was reached. On the other hand, if farm wages were high and land was cheap, it might pay him better to use more land and less labor, cultivating less intensively. Similarly, a manufacturer would substitute machinery for labor if the marginal value of the product he could obtain by using machinery was greater than that of the labor which it allowed him to eliminate. Low wages, on the other hand, might deter technological improvement by making it more expensive to buy a machine than to pay wages for handwork.

Is Marginalism Relevant?

All this is very ingenious, but how relevant is it to the real economic world? The extreme mechanistic abstraction of the school raised doubts. Picture a befuddled consumer, making out an expenditure budget on a form far more intricate than an income-tax blank, drawing up schedules of preference for all his wants, deciding on the last unit of each product he would buy at any given price. Of course no consumer ever does such a thing. Even businessmen make no deliberate marginal calculations; their records contain no equations and graphs of the sort found in the economic textbooks.

The theorists reply that they never pretended to be representing the actual behavior of real persons; this is what people *would* do if they knew their own best interests and were seeking to maximize utility and minimize disutility. The abstract economic man in an abstract world of markets is said to be useful in accounting for general tendencies. But how useful is he? Modern psychology does not regard pleasure and pain quite so naïvely. Do not purchases depend on social custom and private habit as much as on calculated choice? Are prices formed in this way rather than on the basis of rules-of-thumb and interrelated price structures? Above all, what good is a hypothesis so formulated that there is virtually no possibility of testing it, since the necessary facts are not available and ar not likely ever to become so?

What can you *do* with a theory like this, except perhaps t use it as an argument for the *status quo?* Carried out to all i implications of general equilibrium, and revealing an alloca tion of resources in such a way as to offer consumers exactl what they want in the order of preference in which they war it, at the lowest possible prices, the theory is little more tha an impressive elaboration of Adam Smith's "invisible hand that leads everyone to serve the best interests of all if only h intelligently pursues his own advantage.

Alfred Marshall's Restatement

In 1890, some forty years after the publication of Joh Stuart Mill's *Principles,* another English economist returne to the grand tradition by attempting a restatement of the who body of economic thought.

Much had happened, both in actual experience and in doc trine, since Mill had written. Socialism had made little head way; British capitalism, dominant throughout the civilize world, was in its full flight of success. Trade unions had grow hours were shorter, and wages were higher. Victorian opt mism colored ideas. Many social problems remained, but th assumption was that they could be solved by good will opera ing in a favorable environment. The old books were no adapted to the changed institutions. Marshall's *Principles c Economics* was prompted by his observation that "each gen eration looks at its own problems in its own way."

Alfred Marshall (1842-1924) was the son of a cashier of th Bank of England. He studied the classics with the idea of be coming an Anglican clergyman, but found mathematics mor to his liking, was graduated from Cambridge with honors, an got a teaching job in his specialty. For a time he expected t be a physicist, but while at Cambridge he acquired a kee interest in human betterment and read the great economist to discover, if he could, how social problems might be solve His interest in the subject was explicitly that of a reforme and throughout his life he thought of economic science as a instrument for the service of mankind rather than merely a description or defense of what existed. After filling sever other posts, including two years as fellow at Oxford, Marsha was appointed professor of political economy at Cambridge i 1885, and remained at the university, lecturing until 1908 an engaging in research from that date until his death in 1924.

Marshall exercised a dominant influence over British eco

nomic thought during his maturity and for some time after his death. He brought to the subject just the necessary combination of talents—capacity for careful, logical thinking, mastery of mathematics, a broad interest in welfare, and ardor for the improvement of human institutions.

Something more than this, however, accounts for Marshall's success. Rather than developing a whole new body of doctrine or making a brilliant innovation, he built on tradition, wove together contributions from various schools, and constructed an intellectual house of many mansions where a large variety of opinions might feel at home. A severely logical Frenchman or a hotly controversial German would never have thought such moderation possible or useful. In economics Marshall exercised the peculiarly British genius for liberal statesmanship which combines apparent incompatibles, makes room for the new without abandoning the old, expects growth through creative improvement rather than by sharp conflict, and manages all this with the utmost tact, good will, and rationality.

Though his *Principles of Economics* was noted for clarity of expression, it was never widely read by the lay public and did not repeat the popular success of Adam Smith or John Stuart Mill. The subject had, in many of its parts, become too intricate and technical for that.

Characteristics of Marshall's Doctrine

Marshall criticized the concept of the "economic man" as too narrow, since it is with real men that we are concerned, but he retained the "economic man" for his analysis of business, where pecuniary measures are the test, no matter what actually motivates the participants. He praised competition but admitted that it had bad features as well as good ones and should be moderated in some circumstances by cooperation or combination. He substituted the term "freedom of industry and enterprise" for the term "competitive system" as a description of our economic order.

The notion of a primitive paradise where everyone had been happy and good he abandoned as untrue to history and an unfair criticism of our order, but he pictured the existing order as subject to improvement which might lead to approximate such a paradise sometime in the far future.

Marshall saw the evils of poverty and wanted to abolish them; he also saw the benefits of private wealth and did not want to abolish that. He did, however, recommend more moderate and tasteful expenditure rather than "conspicuous consumption," and favored the dedication of wealth to the

public good. He subordinated the concept of land or capita
as material productive agencies to the idea that the organize
or manager who makes good use of all productive factors :
the central factor. The state he also saw as a useful elemer
in economic affairs.

The doctrine of production favored by Marshall admitte
the idea of Malthus that product from land is subject to d
minishing returns, but he added that the collective skill of ma
brings increasing returns which may overbalance the nig
gardliness of nature, and probably will do so. The tendenc
was to produce a surplus above bare subsistence needs; th
nature and distribution of this surplus was the emerging prob
lem.

For the subsistence level of wages Marshall substituted th
"standard of comfort," which he said both increased eff
ciency of production and was made possible by increase
efficiency. He favored trade unions and collective bargainin
No hard and fast law of wages could be true, he thought, sin
wages were the resultant of many forces. He did not believ
that employers and employees, capitalists and landholder
were in fundamental opposition to one another, since they a
had a mutual interest in efficient cooperation.

Marshall made full use of the marginal-utility theories an
wove them all together into a logical structure, with improv
ments of his own. At the same time, as a mathematician, h
warned against the use of mathematics too far separated b
complexity from relatively simple propositions. To the anal
sis of demand and supply under fixed conditions he adde
that though this may explain what happens in the short ru
one must also have a theory of equilibrium for the long ru
when employers are expanding their "fixed" capital with ne
buildings and machinery.

But Marshall did not stop with this price and market anal
sis. He had much to say about "the national dividend"-
what we now call the national income—about how the "di
tributive shares" in it were determined and how their sever
proportions might be changed. His final attitude toward r
form was that while it was desirable, most reformers were
too great a hurry and overestimated the possible speed
change because they expected too much of human nature.
their views were right about the innate chivalry of man, eve
private enterprise would work well, and private proper
would become "harmless at the same time that it becam
unnecessary."

All this and much more offered something to many do
trines and technical specialties in a climate of judicious un

ersalism where one could find a place for almost every idea
nd an exception to almost every rule. The book left many
oors open to wider vistas and new pilgrimages. Logic was
ere but did not dominate; warmth of feeling was moderated
y skepticism. Great variety was placed in an ordered frame-
ork. The whole structure was, in intellectual terms, much
ke the British Victorian realm, where an adored monarch
as combined with democracy and progress, and a peculiarly
sular nation stood at the heart of a many-hued empire on
hich the sun never set.

alue and Prices Again

The tortured theory of value and prices had always been
e heart of classical doctrine. The purists virtually confined
emselves to speculation concerning these matters; econo-
ists of broader scope added many other ideas but never
mitted full discussion of price theory.

Marshall himself made most of his original contributions
this field. For example, he introduced the important con-
ept of elasticity of demand—that is, the degree to which
hanges in price of a given article affect the demand for it.
ery different considerations may influence pricing and pro-
uction policy when demand is highly elastic than when it
less so. By a sort of inversion of Marxist surplus value he
lso spoke of the "consumer's surplus"—that is, the difference
etween what a consumer would be willing to pay for some
rticle if necessary—that is, its real value to him—and what
e actually could buy it for.

Most price theory had been developed on the assumption
at perfect competition prevailed. Later theorists had no diffi-
ulty in working out how prices would be set by a monopolist
ho had knowledge of what would benefit him most: he
ight gain by reducing prices because of increased sales, but
e would stop somewhere short of the point to which compe-
tion presumably would have forced him. All this could be
own neatly by equations and curves. But it had long since
ecome obvious that in selling and buying neither competi-
on of the sort postulated by the economists nor simple
onopoly was characteristic of our order, or was ever likely
become so. Most actual situations were somewhere between
ese extremes. Would it not be possible to use the existing
ols of analysis to parallel more closely the observed reality?

Many of the new classicists who followed Marshall con-
entrated on this problem. They began to think and write
bout "imperfect competition," "monopolistic competition"

(monopolies of differing but competitive products, like brand name foods), "oligopoly" (a few sellers), and "oligopsony" (a few buyers). Here was a glorious new opportunity to spin logic or to fill pages with mathematical formulas. The discussion, in which names of living economists like the English Joan Robinson and the American Edward Chamberlin are prominent, still continues; one can find learned articles on it in almost any current professional economic journal.

Most of this theory, like the classical tradition from which it is derived, does not rest on careful case studies of how businessmen actually do set prices, but on how they would set prices under assumed situations if they were motivated by rational considerations rather easily imagined. Such simplification is necessary for logical deduction. How illuminating it is about what really happens to prices in any actual market under the multifarious forces that affect it, nobody can be sure. Few of the formulas are so conceived that they can be tested against experience.

The price economists do not controvert as a rule the popular belief that the more competition, the better for the consumer. Even this conclusion, however, is not necessarily true under all circumstances. Cases are suspected, and may be frequent, in which a large producer, with lower costs than his small competitors, fails to make price reductions that would increase his sales revenue because he might drive the small fry out of business and so incur public disapproval as a monopoly!

7

The Revolution of
John Maynard Keynes

The urbane complacency of the orthodox economic tradition was roughly jolted by World War I (1914-1918). The war disrupted international trade and payments, led governments to carry out hitherto undreamed-of intervention in economic affairs in order to concentrate production on war necessities, and created huge governmental debts.

No such convulsion had been experienced since the Napoleonic Wars at the beginning of the nineteenth century, and even they had not absorbed nearly so large a portion of the nations' energies or drawn so great a part of their populations into the armed services. Furthermore, the earlier general European war had come near the beginning of the great outburst of industrial expansion rather than after a large part of the civilized world had become industrialized.

Britain after World War I, like other European countries, faced perils hitherto unknown to her. A large share of her overseas investments had been sold to pay for war supplies. She had assumed great governmental debts to the United States. Inflation, with a rapid rise of prices, had occurred during and immediately after the war in almost all countries; it was succeeded by a depression and exceedingly sharp fall of prices in 1920-21. There followed, in the 1920s, a sagging of Britain's export markets, accompanied by persistent unemployment in the export industries.

The trade-union movement had gained great strength; industrial labor could no longer be ignored. In Britain and

Europe it was the core of political parties inspired by sociali[st] philosophy; these parties, though minorities, exercised mu[ch] power and seemed to be growing. A revolution, led by do[c]trinaire Marxists, had actually occurred in Russia and threa[t]ened to spread. Advocates of capitalism felt the ground sha[k]ing under their feet and began to assume the defensive again[st] external attacks.

It was inevitable in this period that economists should [re]examine their theoretical positions and try to discover polici[es] that would restore stability. Even those who placed chi[ef] reliance on the orthodox doctrine that wholesome equilibriu[m] would automatically result if private enterprise and free ma[r]kets were left to work out their own destinies had to adm[it] that since governments had been forced to intervene dras[ti]cally, additional governmental action might be required [to] make possible a new start.

When, in 1929, one of the deepest and longest depressio[ns] that the capitalist world had ever experienced engulfed t[he] United States as well as nations already suffering from t[he] shock of war, the turmoil in thinking was intensified. Anyo[ne] who had a remedy to suggest could get a hearing. Tho[se] writers on economics who, stimulated by the urgent pressur[e] of the times, seemed to have something pertinent to say abo[ut] practical social problems achieved influence. Rarely since t[he] days preceding the French Revolution had the winds of do[c]trine whirled more fiercely.

Two key areas of technical economics came to the fore [in] these turbulent years. One was the theory of money, sin[ce] money apparently had so much to do with inflation and d[e]flation, accompanied by wide swings of prices, as well as wi[th] the problems of foreign exchange and international paymen[ts]. The other was the question whether national governmen[ts] could and should do anything to diminish unemployment, a[nd] if so, what.

Many were concerned in these discussions, but it so ha[p]pened that both key problems received what was probably th[e] most systematic and most elegant theoretical treatment in t[he] works of John Maynard Keynes, an economist connected wi[th] Cambridge University, and a favorite student of the reigni[ng] Marshall. Though the sort of measures he favored had be[en] proposed by others, most of those who urged them had com[e] from the "underworld" of economics; their conclusions we[re] not supported by the scholarly reasoning known only to t[he] initiate. But Keynes, using the approved orthodox metho[ds] and speaking from the very heart of academic respectabilit[y,] came out with conclusions that denied beliefs held by mo[st] orthodox economists for more than a century. Here was a revolution indeed!

Keynes thus came to be the symbol of a whole new style of economic doctrine. Before inspecting his theory it will be well to look briefly at some of the ideas which preceded him.

Ideas about Money—the Gold Standard

Theoretical discussion of money runs back as far as thought about any economic subject. Does "good" money have an "intrinsic value" or is its value the result of accepted use as a medium of exchange? Are gold and silver commodities, like other goods, or are they chiefly common measures of the values of other commodities? Is money productive, or is it sterile? Is it capital, or is it merely the means with which capital and other goods are bought, and the measure of their value? We need not concern ourselves with these and similar questions, to which no demonstrable answer can be given.

It is important, however, to recognize broadly the historical development of money. When the early classical economists wrote, money was thought of as the precious metals and the coins minted from them. People also used pieces of paper in buying and selling, in borrowing and lending, but these written or printed slips were usually promises to pay gold or silver, either on demand or at some specified time—in other words, notes. Both governments and private persons issued such notes. Sometimes people used certificates indicating that metal money was on deposit and could be had by presenting the paper. Sometimes they used even warehouse certificates, indicating the ownership of goods supposed to be worth a certain amount of money. Few persons, however, questioned that the only "real" money, and the basis for all other media of exchange, consisted of precious metals of specified purity often called specie).

As banking became more prevalent, bank notes began to circulate more and more widely instead of actual coins. A bank note represented a promise by a bank to pay specie on demand. When, however, a bank had the reputation of redeeming its notes whenever they were presented for payment, most recipients of the notes would not bother to redeem them but would simply pass them on as payments to other persons. Banks could, and did, issue many more notes than they held cash to redeem. A bank was thought to be sound if it had on hand a specie reserve which was a mere fraction of its promises to pay, provided its total assets did not fall short of its liabilities. Thus, coin money was augmented by a muc[h] larger quantity of paper money.

The next means of expanding the means of payment oc[-] curred through the practice of transferring bank deposits b[y] check. This practice was not generally used in the Unite[d] States until after the Civil War, but it now accounts for by fa[r] the greatest volume of payments.

A bank, for example, makes a loan to a borrower by merel[y] writing figures on his deposit account—in exchange for whic[h] the borrower gives the bank a promise to repay. The borrowe[r] then uses his loan-created deposit by sending checks to hi[s] creditors, who in turn deposit the checks to their account[s.] The banking system records the whole affair by subtracting o[r] adding figures to depositors' accounts. Neither any coin n[or] any paper money has changed hands in this kind of transac[-] tion. Bank deposits, consisting of figures on books, have b[e-] come money, in the functional sense.

Under these circumstances the idea that the value of a[ll] kinds of money is fundamentally sustained by the privileg[e] of exchanging it for an equal value of gold or silver rests on [a] rather slender basis, but this belief was almost universally hel[d] until the 1930s.

The Quantity Theory of Money

Adam Smith, like most economists of his day, believed tha[t] the flooding of gold and silver into Europe from the Spanis[h] possessions in America had caused the long-continued rise i[n] prices which had been observed since the beginning of th[e] sixteenth century. Economists generally accepted the quan[-] tity theory of money—the more money, the higher are price[s;] the less money, the lower are prices.

Though this idea was picked to pieces and discussed ex[-] tensively by economic writers for generations, it persisted i[n] some form or other and was widely accepted not only b[y] theorists but by the general public.

A standard grievance of the American colonists was tha[t] they did not have enough money to finance their businesse[s] and pay their debts. When they expanded the supply by print[-] ing bills or notes, the British government stopped them t[o] avoid inflation of the currency, rapid rise in prices, and injur[y] to the creditors. When John Law in France tried to remedy [a] currency shortage by issuing paper money based on land, th[e] resulting overissue resulted in inflation followed by collaps[e.] Fiat paper money (unbacked by gold) issued during th[e]

American Revolutionary War became so plentiful that it was almost worthless. Political controversies in the United States raged during the whole nineteenth century about kinds and quantities of money. Though the question was confused by the idea that when dollars buy less because of high prices they had less value merely because bills were irredeemable in specie, the feeling that the quantity in circulation had much to do with the purchasing power of the dollar was also prominent.

Irving Fisher (1867-1947), professor of economics at Yale in the early twentieth century, concentrated on monetary problems with the hope of stabilizing prices and so moderating both booms and depressions. He, and those who agreed with him, reasoned as follows. Rising prices are the result of too much money; falling prices of too little. Everybody agrees that extreme inflation, when prices rise so high that money becomes virtually worthless, and panics, when loans become temporarily unobtainable, should be prevented. But it is important also to avoid more moderate fluctuations of the general price level, since rising prices encourage everybody to buy in the hope of speculative profits and falling prices lead everybody to sell, or refrain from buying. These alternate waves of buying and selling intensify the irregularity of business activity and employment.

What is "too much" or "too little" money? The practical test was, Fisher thought, what happened to prices; rising prices should lead to restriction of the money supply, falling prices to its expansion. Therefore the "stable-money" advocates spent much laborious study on statistical measures of prices, to obtain what they could regard as a representative index of price tendencies.

Their version of the quantity theory was refined and given more logical precision. Money, regarded as a medium of exchange, must be defined as including everything used for that purpose—not merely gold, silver, and paper money but also bank deposits. Whenever an exchange of goods or services for dollars takes place, the exchange constitutes use of money, or a "transaction." Now, if the circulation of money changes in the same proportion as the number of transactions, the level of average prices must remain the same. If circulation increases more rapidly than transactions, prices will rise, and vice versa. Circulation, in turn, results both from the *amount* of money in use and from the *velocity* with which it passes from hand to hand.

All this was mathematically expressed in Irving Fisher's

celebrated equation of exchange, the simplest form of which is, as first written by Simon Newcomb,

$$M V = P T$$

where, at any given time, M is the quantity of money, V the velocity of its circulation, P the general price level, and T the number of transactions.

In the eyes of the stable-money theorists, M was the controllable factor of this equation and P the factor which it was important to keep constant. Therefore deliberate changes in M ought to be made to counterbalance any changes that might occur in V and T. Suppose, for example, that as the country grew, the number of transactions doubled while the velocity of circulation remained the same. In that case, thought Fisher the quantity of money ought also to be doubled. The equation would thus become

$$(2 M) V = P (2 T),$$ and no change in prices would occur.

If the quantity of money and the velocity of its circulation should remain the same while transactions doubled, the result would inevitably be

$$M V = (½ P) (2 T),$$ prices being cut in half.

The mathematically inclined reader can devise such variations at his pleasure.

All this may seem innocent enough to the unsuspecting layman, but at the point where Fisher began to advocate deliberate management of the quantity of money he ran smack into conservative orthodoxy. The amount of bank money was supposed to be dependent in the first instance on the free play of profit-seeking among competitive banks and those who borrowed from them, and ultimately on the gold reserve. As for government-issued paper money, the government's only duty was to keep it "sound" by making sure that every Treasury note or certificate could be exchanged for a permanently fixed weight of monetary metal. Yet Fisher wanted to vary that weight as prices went up or down, so that a dollar, instead of buying an invariable amount of gold, would, he thought buy an invariable amount of commodities.

He assumed, in accordance with classical reasoning, that to make the dollar worth less or more in gold would automatically change the quantity of money in circulation. But his suggested snackling of the "invisible hand," or natural law, which was supposed to arrange all things for the best if only there were no interference, was thought by the orthodox highly dangerous, not to say impious.

The quantity theory of money has been subject to much revision; and the role of gold in determining quantity of money has become highly suspect even when the gold standard is retained. Yet management of the circulation of money in the public interest is now believed desirable by conservatives themselves, even though it is no longer regarded as the sole means of taming the business cycle. Irving Fisher's equation of exchange has its uses, though it is not a comprehensive or accurate statement of the causes of inflation and depression.

Money in Foreign Trade

Beginning as early as Thomas Mun, economists spelled out the theory of foreign exchange and its relationship to international payments. The subsequent classical tradition applied to this realm of thinking, as to all others with which it dealt, the doctrine that a healthy equilibrium would tend to prevail if only no one interfered with the "natural and simple system of liberty." It was supposed to work in this way:

Virtually every nation was on the gold standard. That meant that any possessor of a unit of its currency could freely exchange it for a certain amount of gold, the gold value of the unit being legally established by each nation for its own money. This situation automatically determined the relative values of the several national units of currency, since each was fixed in terms of a common unit—gold.

Now suppose that payments by, say, Englishmen to Americans persistently ran larger than payments by Americans to Englishmen. That would, in the foreign-exchange markets, increase the demand for dollars and decrease that for pounds, so that the exchange value of the pound would tend to sag. The inevitable result would be a flow of gold from London to New York, because those who wanted to buy dollars with pounds would not pay an increased number of pounds for them as long as they could buy gold in England with their pounds at the fixed legal rate and pay their debts in the United States with gold.

The United States, into which gold was flowing, would have more money as a result and American prices would rise. Britain, losing gold, would have less money and so British prices would fall. In consequence of this change in relative prices, Americans would buy more British goods, while the British would buy fewer American goods. Americans would therefore have to make more payments in pounds than before, while Britishers were making fewer payments in dollars. This change in demand for and supply of the two currencies would stop the tendency of the pound to fall. It would thus end the outflow of gold from the United Kingdom. The spread between prices in the two countries would then cease to widen. Trade and payments would again be in equilibrium at a "natural" level of comparative prices, and all would be for the best in the best of all possible worlds.

This theory, beautifully logical though it is, is open to numerous objections as a description of what happens. The connection between gold and the quantity of money, and between the quantity of money and prices, is not so close as the theory implies. The need of one nation to buy from or make payments to another is not nearly so responsive to change in relative prices as the theory supposes. Yet the doctrine was regarded by most authorities in the 1920s not only as a pleasing abstraction but as a practical guide to policy.

It led to the attempt to put all nations back on the gold standard in the mid-1920s—the use of gold in international payments having been suspended during the war. But the whole structure of international payments had been suddenly altered by the international war debts, to mention only one important change. The English pound, because of shrinking exports and outgoing war-debt payments, now had a tendency to fall.

Though Britain lost gold and the United States gained it, commodity prices did not rise in the United States. Britain was in danger of losing its gold reserve. At this outcome British financial authorities were highly indignant and charged the United States with "sterilizing" gold to keep its price level down, contrary to "natural law." (This sterilization took place not by changing the gold content of the dollar, as Fisher had recommended, but supposedly by central banking policy which prevented the issue of as much money as the gold reserve would have allowed.)

Thus the newly favored policy of monetary management was attacked on the ground that it endangered British recovery. It is extremely doubtful that the stability of American commodity prices in the 1920s actually did result from banking policy, but the incident underlined the dilemma—if each separate nation was to manage its money, what would happen to international stability?

Money and Interest Rates

Interest is the price charged for borrowed money; therefore the orthodox economists, with their eyes focused on the theory of prices, had much to say about interest. They had of course applied to it the obvious doctrine of supply and demand, and from there had gone on into endless elaborations as to what influences the supply of, and the demand for, loan funds.

Turgot pointed out that savings increased the supply. Say stated that relative risk, and ability quickly to turn debts into ready money (liquidity), were factors in rates charged for loans. The marginalists introduced their thinking into the subject, alleging that demand for the "last unit" of money which the borrower wanted to borrow and the lender was willing to lend expressed the rate. Some emphasized that interest was the reward for those who were willing to undergo abstinence in order to save, and therefore that high interest rates, as a reward for waiting, encouraged saving and increased the supply of capital. At the same time higher rates would reduce the willingness of borrowers to borrow.

From all these and other ideas, the classicists wove together in the realm of interest their customary equilibrium theory. There was a "natural" rate at which all relevant factors would be in equilibrium. Free enterprise allowed this rate to be approximated. It was the rate that would prompt just the right amount of saving to satisfy the legitimate need for money by those who wanted to use it as capital, that is, for production. Of course demand for, and supply of, money at the "natural" rate would be equal.

Since the steady growth of capital was responsible for material progress, anyone who should try to keep the rate of interest above or below the "natural" rate resulting from supply and demand would be laying profane hands on the economic holy of holies. He would be upsetting the delicate mechanism by which savings were attracted and by which the need of society for new capital was met. Yet that is just what the innovators proposed. One of the instruments which money-managers were now expected to use, for example, was to lower the bank rate for loans in order to encourage more use of credit during depressions, and to raise the rate during booms in order to discourage unhealthy expansion.

The still more daring proposals of Keynes in this realm outraged all the high priests of the classical dogma.

Welfare Economists

The other main stream of thought which influenced the type of analysis later brilliantly made by Keynes was the idea that government itself could and should do something to minimize unemployment, which experience had shown frequently characterized the operation of systems of private enterprise.

Contributors to this stream included economists in the classical tradition like A. C. Pigou (1877–   ), Marshall's successor at Cambridge University, whose *The Economics of Welfare* was published in 1920. Pigou arranged all his theoretical com ments about the question: What policy or policies would mos benefit the majority of the population?

In this framework Pigou made prominent use of the idea o the national income—that is, the total income of everybody i a nation. This sum he called by Marshall's term, "the nationa dividend." Reliable figures on the basis of which it could be estimated had been scanty before 1914, but the welfare econ omists contributed much to the theory of the methods to be used in estimating it. They discussed the best means of increas ing it, and the possibility of achieving a nearer approach to equality in its distribution—subjects that had interested many theorists in the past. In addition, they devoted great attention to the successive upward and downward waves of aggregate income (the trade cycle or business cycle), a subject the ortho dox had been inclined to slight. Pigou concluded that there was much to be said for the suggestion that public works ought to be increased in depressions and decreased in booms. He was wholeheartedly in favor of a national system of labor ex changes (employment agencies) and of national social insur ance.

Other welfare economists were outright opponents rather than modifiers of the classical system of thought. Prominent among these was John A. Hobson (1858-1940), who was grad uated from Oxford but devoted himself to reform rather than to achieving academic reputation. Hobson, in his *Work and Wealth* (1914) and many other books and articles, argued that accepted theory and practice laid too much stress on the mere quantity of production without relation to the human costs to individual workers and society, and without sufficient consid eration of the welfare of the consumer. The pecuniary standard of values, he believed, did not necessarily measure maximum welfare. He objected, too, to the mechanical method of the orthodox theorists, which falsely "treated every human action as a means to the production of non-humanly valued wealth."

Hobson wanted a partially socialized system which would de-emphasize the incentive of monetary gain except where that motive would be socially useful, as in new, experimental, o competitive industries. He also was an ardent advocate of gov ernmental spending to smooth out the waves of private pro duction. The economic theory on the basis of which he did so was scoffed at by orthodox economists since he attributed cyclical unemployment to the production of more than con sumers could buy—a development classical theory regarded as impossible.

The use of public works to relieve unemployment is an old suggestion—it has been proposed, and even tried, time after time as far back as recorded history goes. It had many supporters long before Pigou, Hobson, or John Maynard Keynes wrote. The lack Keynes supplied was not the need for some new and ingenious scheme so much as a theoretical rationalization which could clothe in academic respectability some of the proposals already prominent, proposals that the pressure of events would probably have forced into action if he had never lived.

A Pre-Keynesian—Wicksell

A Swedish economist, whose works were not translated into English, until late in his career and hence had little influence on thinking in America and England until recently, nevertheless was an important predecessor of the Keynesian type of thinking; Keynes, who did read him, acknowledged the debt.

John Gustav Knut Wicksell (1851–1926) began his public life as a social and political reformer. He was by training and profession a mathematician, and later turned to economics to learn how his social aims might be carried out rather than to justify existing conditions. In this he differed radically with most of his contemporaries of the continental marginal school like Böhm-Bawerk and Menger, who conceived their work as defenses of capitalist, against Marxist, value theory.

With the details of his voluminous writing we cannot deal without going into technical mazes, but his main contributions are rather simple.

Wicksell concentrated attention on the changes of a real world over time, with its cycles of generally rising and generally falling prices, rather than on the imaginary static world where everything was supposed to work itself out for the best. Why, he asked, do all prices tend to rise or fall at the same time?

He answered that aggregate demand may for a while be greater or less than aggregate supply. This answer contradicted Say's "law" that there can be no such thing as general overproduction.

Wicksell's answer to the question of why such changes occur introduced analysis of the flow of the national income. Add up the incomes of everyone in a nation and your total shows what the people of that nation can spend, i.e., their potential aggregate demand. The part of incomes spent for consumption obviously creates demand. What is not spent for consumption, that is, what is saved, may become a part of demand if it is used by somebody who puts the money into houses, factories, or machinery. But more (or less) may be saved than is actually invested in such capital goods. When saving is greater than investment, money is held out of the income-spending flow, demand will decline and prices will fall. When saving is less than investment, total demand will increase and prices will rise. This same idea is the heart of the Keynes theory of employment.

Wicksell made the transition from a quantitive theory of mere *money*, to a theory of *income*. He urged deliberate variation of the interest rate to help maintain an equilibrium between saving and investment. He thus pioneered in bringing the theory of money into the same universe of discourse with the theory of general price movements and the theory of national income.

Wicksell stimulated a number of well-known living Swedish economists, for example, Myrdal, Lundberg, and Ohlin, to carry similar work further. They became known as the Stockholm School. Their doctrine, though parallel to that of Keynes, differs in detail and presents a less comprehensive theoretical system. One of their principal contributions is an analysis of how the expectations of businessmen and investors differ from what actually happens if they act on the basis of their expectations. In technical jargon, this is known as *ex ante* and *ex post* analysis.

Keynes Himself

John Maynard Keynes (1883-1946) was the son of a minor economist of the classical school, John Neville Keynes. His mother, a graduate of Cambridge University, ultimately became lord mayor of the city of Cambridge. He was a brilliant student in mathematics, politics, and philosophy.

In order to prepare for a Civil Service examination and enter a career as a government official Keynes studied economics with Marshall. Passing second in the examination, he was appointed to the India Office. There, where he jokingly said his principal duty was to read the newspapers, he spent much time working on a dissertation in economics which he hoped would bring him an appointment as a Cambridge fellow. The appointment went to another, but Marshall saw promise in him, offered him a lectureship, and paid the stipend out of his own pocket. Keynes revised the thesis and the next year received his fellowship.

While on the Cambridge faculty Keynes became editor of the *Economic Journal*, largely because of Marshall's backing.

and devoted himself to study of monetary theory. In 1913 he was appointed member of a Royal Commission on Indian Currency and Finance, where he distinguished himself by a masterly memorandum combining theoretical knowledge with practical shrewdness. Keynes thereafter continued to think of theory as an instrument to inform practice in meeting specific problems. If a new policy seemed desirable on pragmatic grounds, he would modify his theory to accord with what he thought ought to be done, even though his chameleon-like changes caused him to be accused of inconsistency or even insincerity. But he had none of the temperament of the rigid doctrinaire.

During World War I Keynes was called into the Treasury, and at its close attended the peace conference as an adviser. The result was his sensationally successful book, *The Economic Consequences of the Peace*, attacking the peace terms because they called for impossibly high reparations from Germany while depriving the defeated nation of the means of making foreign payments, and because they shattered the economic unity of a hitherto reasonably productive Europe. This attitude brought bitter attacks as well as wide support. The United States, after World War II, has devoted great effort and huge sums of money trying to avoid or even to undo the mistakes Keynes saw in the earlier settlement.

The book also revealed the young economist as a master of literary style, though he could be difficult enough when he wanted to impress fellow economists. His brief character sketches of the "big four"—Wilson, Lloyd George, Clemenceau, and Orlando—whom he despised for making stupid blunders, could scarcely be surpassed by the most skillful writer of fiction in biting evocation of character.

Keynes now became a man of many interests. He retained his editorship of the *Economic Journal* and his connection with Cambridge, which he visited during long week-ends to lecture and teach. At the same time he acted as chairman of an important insurance company and managed an investment company in which he made money both for himself and his college. He bought the great British weekly, the *Nation*, merged it with the *New Statesman*, wrote books, discussed public affairs in the periodicals he owned, married a charming dancer of the Russian ballet, Lydia Lopokowa, subsidized the ballet and collected paintings. He had also written a treatise or two on philosophy. His success as a businessman and his aesthetic tastes were reminiscent of Ricardo, though he had come a long distance from the old master in economic thinking.

It was in 1925 that England, under the guidance of Winston

Churchill as Chancellor of the Exchequer, returned to th
gold standard at the prewar gold value of the pound. Th
result, Keynes foresaw, and cogently said, would be intensif
cation of depression and unemployment in the export indu:
tries, as well as a shortage of gold reserves for the bankin
system. Classical theory prescribed for these ills a fall of Britis
prices and wages brought by the automatic action of the gol
standard; the deflation would cure depression by reducin
the prices and thus increasing the sale of British export good:
If labor would not accept the required wage reductions,
would be responsible for its own unemployment.

Keynes was roused to assault on both the policy and th
supporting theory. Why should British labor be made to er
dure misery for an abstraction like the gold standard? In an
case it would not voluntarily submit. Much better to devalu
the pound than to try to reach a precarious balance throug
the route of painful price adjustments. This episode set Keyne
to revising monetary theory and the theory of wages; his pos:
tion on it formed a precedent for his doctrine that nation
monetary management was necessary even though it migh
lead to embarrassing international consequences.

Keynes settled down to serious theoretical work in the lat
1920s; he gave a series of lectures at Cambridge on the theor
of money; he was a member of the Macmillan Committee o
Finance and Industry to recommend policy for dealing wit
the long-continued decline of the export industries. There h
gave vigorous expression to his unorthodox views. Soon h
tried to formulate his theories systematically in a forbiddin
two-volume *Treatise on Money* (1930). The work showed th
direction of his thought, but it was faulty for several reasons
among them that the assumptions he adopted for purposes c
simplicity were too far from reality. Unsatisfied, Keynes wer
on reading, thinking, discussing with his colleagues.

The world went from bad to worse during the Great De
pression beginning in 1929. Keynes finally formulated his sys
tem with satisfaction to himself in his major work, *The Gen
eral Theory of Employment, Interest and Money*. It was pub
lished in 1936, when the New Deal was in full course in th
United States; unemployment, though diminished, had no
been conquered. No scholarly document since Adam Smith
Wealth of Nations had seemed to fit more closely the need o
the times. Few except academic economists read it or coul
understand it if they tried, but to many theorists in the classi
cal tradition it provided a marvelous rationalization for pol
cies they thought necessary but could not hitherto embrac
with a clear conscience.

Keynes had visited the United States, and during the administration of Franklin D. Roosevelt came again to consult on recovery policy. During World War II he served the coalition government of the United Kingdom, as did virtually all British economists, advising on the complicated problems of taxation, finance, mobilization of the economy, and postwar planning. By this time he had been honored by the government with a peerage. After the war he journeyed to the United States again and negotiated the agreement for the British credit which preceded the Marshall Plan. The agreed sum was not large enough to see Britain through, and he knew it, yet he had to battle for its acceptance by a reluctant House of Lords. Shortly thereafter he succumbed to heart trouble, from which he had been a chronic sufferer for some years.

Keynes' General Theory—Employment

Keynes began his great book with an adroit introduction. The classical and neo-classical doctrines were logical enough, he said, but they could hold true only in a special case—the case limited by the many assumptions, expressed or implied, that the classicists adopted as a basis of their hypothetical reasoning. This case was not the actual one. He proposed to develop a general theory for which unrealistic assumptions were unnecessary.

Classical theory had little to say about the fluctuations of employment. It assumed that production in any free-enterprise economy would fluctuate closely about a balance between demand and supply, and that in this equilibristic elysium everyone who was worth hiring could get a job. Wages, like any other price, would fall if the supply of labor exceeded the demand for it. As soon as labor became cheap enough, employers would find it profitable to hire more. The implication was that persistent or widespread unemployment could be caused only by stubborn refusal of workers to accept low enough wages; there was, by and large, no such thing as involuntary unemployment.

Keynes saw that this theory, however logical, was inapplicable to the real world. Workers did habitually refuse to accept wage reductions, even if they had to quit their jobs; trade unions were just as real as any other part of the economic system. Widespread unemployment was, in fact, frequent, and the experience of many workers during depressions was that they could not get a job at any wage at all.

Keynes was too good a student of classical models not to substitute an equilibrium theory of his own. There could be an equilibrium between demand and supply, he argued, which would tend to set the level of employment, but this level might be so low as to leave a large number of competent and willing workers unemployed. Keynes argued that high wages could not be the chief cause of unemployment. He believed that a wage cut was theoretically equivalent in its effect to a fall in the rate of interest, and that anything which might be accomplished by wage reduction could be better achieved by lower interest rates, though neither of these could be a sufficient remedy.

If general unemployment was not due solely to too high a wage level, what did cause it? Here Keynes applied a truism, all the implications of which had not been recognized by the classicists. Every payment has two sides. The recipient receives as much income as the payer spends—just that much and no more. Income flows in a circle about the economic system. If there is a fall in general demand, it must be due to a shrinkage in the flow of income. One must look for the cause of the shrinkage. Who is holding money out of the stream; why does he do it?

Clearly, money spent by consumers (and everybody is a consumer) is not withheld from spending. Any money withheld, at least temporarily, is what people save. The classicists usually assumed that all savings were necessarily spent by those who invest in manufacture or other business. Keynes, like Wicksell, challenged this assumption. The savers are often not the same people as the investors and have different motives. You and I pay money into a savings bank or an insurance company; the money is not actually spent unless somebody wants to use it to put up houses or enlarge a business. More (or less) money may be withheld from consumption than is being spent for investment purposes; this is what causes downward or upward swings in the stream of aggregate income.

Keynes of course put this reasoning into mathematical form, in the true classical tradition. His fundamental equation is

$$Y = C + I$$

Y being aggregate income; C being the amount spent for goods for consumers; I being the amount spent for, and consequently received in the production of, investment goods (capital).

It is obvious too that savings (S) is aggregate income (Y) less the amount spent for consumption (C). Or

$$S = Y - C$$

Here are two simultaneous equations from which it is easy to derive $S = I$, or savings equal investment. But is it not Keynes' principal contention that savings do not necessarily equal investment? Are not these equations therefore inconsistent with his theory? The answer is that people sometimes *intend* to save more than is being invested, but if the actual investment does not equal what people intend to save, their incomes will fall, so they will in reality save no more than is being invested. When *intended* savings equal *intended* investment, the economy is in equilibrium. The equations therefore became a picture of balance between intended investment and intended savings, toward which the economy is supposed to tend. If the economy should actually remain in equilibrium, no change would occur in either savings or investment.

Keynes and the Theory of Money

The classicists held that the money market, like any other market, tended toward an equilibrium of demand and supply at a "natural" rate of interest. A higher rate of interest would increase the supply by encouraging saving and would decrease the demand by discouraging borrowing; a lower rate would do the contrary. Since those who borrowed money were mainly the people who wanted it for investment, the classical equilibrium theory implied that there was no problem in equating savings and investment.

Keynes of course challenged this theory in declaring that savings and investment were independently determined. As in the case of wages, he minimized the role of prices in affecting demand and supply. Other factors, he thought, were more important.

Here he differed not only with the classical writers but with the advocates of monetary management like Fisher or Wicksell, who based their policy recommendations on the classical quantity theory of money. Wicksell, for example, proposed as a remedy for depression a reduction of the interest rate to induce more investment and less saving. Keynes did not oppose this policy, but he believed it insufficient. He thought variations in the interest rate had almost nothing to do with the amount saved and were far from wholly responsible for the amount invested.

The Keynesian Determinants of Saving

How do people decide how to divide their incomes between consumption and saving? Keynes thought this largely a matter of habit and necessity. People acquire a certain standard of living and spend what is necessary to maintain it. The rest of their incomes, if there is any left, they save, without much regard to the rate of interest they can earn. The amount spent at a given level of income, an amount resulting from a combination of psychological and social forces, Keynes called the "propensity to consume." Since this amount is an important factor in mathematical expressions, it is also called the "consumption function."

What is saved is merely what is not spent on consumer goods. In fractions or percentages, it is the complement of consumption. That is, if consumption is four-fifths, or 80 per cent, of income, saving will be one-fifth, or 20 per cent, of income.

Now we come to a critical part of the Keynes doctrine. How much those in any income bracket will save, he believed, varies with the size of their income. Poor people have to spend all they make; they save little or nothing. Rich people have a much bigger surplus; consequently their savings propensity runs higher. This observation is in accord with common sense and has been corroborated by statistics. Keynes drew important inferences from it bearing on his main theory.

One inference is that in a society where there is little change either in total income or its distribution, there is likely to be little change in total savings. Since consumption will be stable, its complement, savings, will be stable also. This conclusion conceals assumptions which may be contrary to reality—for example, that *nothing* ever causes a change in spending for consumer goods except a change in the size of income.

By the aid of these conclusions he was able to introduce into his theory intricate marginal calculations of the traditional sort. What determines changes in saving, he held, depends on the changes in propensity to consume resulting from marginal increments to income.

The Keynesian Determinants of Investment

Keynes held that though the consumption function which determines savings is relatively stable, investment is relatively unstable. It depends largely on the anticipation of business men about what they can sell in the future. If they expect a contracting market, they will not invest anything, no matter how cheaply they may be able to get the necessary money. They may actually disinvest, and habitually try to do so in managing their inventories. If, on the contrary, businessmen expect larger sales and higher prices, they will cheerfully invest large sums even if they have to pay high rates for the money.

Here also Keynes introduced the marginal conception. A

given new unit of capital equipment is expected to yield over a period of years a certain return on the outlay which the businessman must make in order to buy it. If this expected return is above the rate of interest, the investment will be made. If it is below, the investment will not be made. The "marginal efficiency" of capital therefore determines whether or not investment will increase. When the marginal efficiency of capital equals the interest rate, investment will be in equilibrium. Business will go on investing what is necessary to produce at the current rate but will not expand.

The important conclusion from the whole theory is: since consumption is relatively stable and investment relatively volatile, variations in investment are largely responsible for variations in national income.

The Multiplier

How much will a given rise in investment enlarge national income? Much more, Keynes argued, than the amount of the new investment itself. Investment exercises a powerful leverage, even though the new investment in any one year is a small slice of the national income.

Suppose a manufacturer spends one million dollars to put up a new plant. The workers he hires to build it and the sellers of the materials or services get the million. It is income for them. If they in turn spend all of it, if the recipients on the third round spend all, and so on indefinitely, the money will keep on generating income at each turnover, forever. The sum of income it can ultimately generate will be infinite.

As a matter of fact, however, there will be "leakage," because on each round of payment and spending some money will be withheld, or saved. This limits the amount of income generated by investment. It also makes possible the calculation of *how much*, provided the savings function is known. As it figures out, the "multiplier" which shows how much a given new investment will augment total income is the reciprocal of the fraction of the marginal income being saved. That is, if one-fourth of the increment in income is being saved, the multiplier will be four, and an investment of one million dollars will ultimately generate an extra income of four million. How long it will take to do this depends on the velocity of income circulation.

One happy result of this calculation is that each new investment brings forth exactly enough saving to pay for itself In our example, the investment of one million generated an income of four million, one-fourth of which, or one million, was saved. This seems almost too good to be true, until we remember that the process is not supposed to work in reverse. Savings do not generate investment. If the new, additional million saved is not invested by somebody, income will cease its growth. General equilibrium will occur when investment is in equilibrium.

The Role of Government

So far government has not entered this theoretical picture. Yet it is a highly important economic factor, becoming more important decade by decade. It receives income; it spends; it saves or invests. Keynes and his followers did not ignore it; indeed, conclusions as to governmental fiscal policy were the major practical outcome of their analysis.

Hitherto the orthodox had supposed the proper role of government in fiscal matters to be neutral. Government had certain inescapable functions which cost money; it should collect the necessary money from taxpayers and pay its bills. If unfortunately it was obliged to incur indebtedness, it ought to pay its debt as soon as possible.

Keynesian doctrine looks upon government as properly a great compensating factor for the uncontrollable vagaries of private capitalism. Government can invest when business anticipations are so poor that business investment slumps, since government does not have to earn a profit. A government investment generates just as much income as a private investment of equal size. The income it generates will enlarge markets and so ought to improve the anticipations of businessmen.

If the government desires to remedy a decline in national income caused by the fact that private savings are tending to pile up faster than new investments, it should spend money which is not taken from the public's income. That is, it should spend more than it is receiving in taxes; it should have a deficit. The money borrowed to meet this deficit should come either from private savings which are being hoarded or from bank credit.

The reverse policy should be followed when inflation is threatened, that is, when full productive capacity has been narrowly approached and major price rises are threatened. When the government spent more than it collected, it was acting as an investor. Now, on the contrary, it should act as a saver. This means collecting more than it spends—having a surplus.

The main stress of the doctrine was thus laid not on direct governmental intervention with the operation of private enterprise, but rather on a compensatory fiscal policy which would moderate the major flaw of the order while leaving it substantially free. Keynes was concerned chiefly with stabilizing the total income; he believed strongly in free markets to determine specific prices and to allocate resources to the best interest of the consumer.

One further aspect of his thinking deserves mention even in so condensed a summary. Wide disparity in the distribution of income, he believed, tended to increase saving and limit investment. A large number of poorly paid people at one end of the scale would not provide enough demand to attract investment of as much as a prosperous class would attempt to save. In this he agreed with Marx, Hobson, and many others whose theories were less skillfully articulated.

Economic health of the nation required more nearly equal distribution. Therefore he favored progressive taxes, combined with various measures of social insurance and public services, to help redistribute income. He also advocated a permanent policy of keeping interest rates low, both because it would help to stimulate investment, and because it would discourage the growth of a rich *rentier* class, living on ownership of securities rather than by production. He foresaw "euthanasia of the *rentier*" by a tendency of interest to fall toward zero.

It is curious that by his intricate process of reasoning Keynes returned almost to the doctrine of interest held by Aristotle and the medieval church. Like the ancients, he thought money in itself unproductive and favored a society in which those who produced goods and services, rather than those who rigged markets and hoarded riches, should receive the rewards.

International Considerations

If authorities in each nation were to be assigned the role of compensating for ups and downs of employment, of controlling their price levels without regard to the gold standard, of redistributing income, and of keeping interest rates low no matter what the "natural" or market rate might be, what would happen to free markets in international trade? Would not the several nations return to mercantilist policies? Would they not strive for separate advantage by competitive depreciation of exchanges, import controls, encouragement of exports by subsidies, and innumerable other interferences with the ideal free flow of world trade and investment which was supposed to embody the classical utopia of international laissez faire, with its division of labor, its international specialization, and supposed automatic equilibrium?

As a result of the depression and World War II, nations were already far on the road back to mercantilist interferences. Keynes was accused of encouraging this tendency. He did, indeed, think that laissez faire was dead internationally as well as within nations. He accepted the consequences of his logic in this realm as in others. But he believed firmly in an international economic community and world-wide trade, and, as in internal affairs, supported new policies and instruments by which to achieve it.

The trade walls which had been erected along international boundaries were largely the result of the depression and the unemployment which had arisen within the chief industrial and trading nations, notably the United States and Britain. Lacking customary export markets, nations had had to restrict imports and push foreign sales. Nations with adverse balances of trade, formerly financed by inflow of new investment from abroad, or by earnings on old investment, suffered shortages of foreign exchange when the flow of investment and earnings ceased. These imbalances had destroyed the gold standard for good. Need for rationing scarce foreign exchange had necessitated exchange controls. The way out was not to begin by abolishing interference with trade and payments but rather to abolish the causes of the interference—unemployment and depression within the great nations on whom the others depended both for foreign markets and for capital.

When faced with a sharp alternative between restoring international laissez faire and pursuing internal policies which he thought would assure employment, Keynes unhesitatingly chose the latter. It would be not only futile but dangerous, he reasoned, to suppress the symptoms without curing the disease. But he set himself the task of bringing about parallel internal policies, on a world-wide scale, which would harmonize the interests of the nations and would end attempts to bulwark internal prosperity by a "beggar-my-neighbor" policy.

Crucial in this effort were the attitudes of the British Commonwealth and the United States. Their joint production and trade bulked so large in the world economy that together they could ease the difficulties for all; others would be sure to follow their lead. If both pursued the same policy of promoting full employment, with due regard for the differing necessities of each, their interests would not clash.

Britain, to maintain employment, would need large markets

or exports and would have to encourage production of export
goods, not by subsidy but by internal allocation of resources.
The United States could stabilize employment without much
aid from exports because of its great internal market; its part
as a creditor nation would be to lower barriers to imports and
to supply new international investment. Full employment in
both countries would shoot high-voltage energy through world-
wide circuits. Trade restrictions could be removed without
peril as production and trade grew, but only if and as they
grew.

Keynes participated in the work which led to the International Monetary Fund to stabilize foreign exchanges, and to
the International Bank for Reconstruction and Development,
formed for the purpose of encouraging international investment. His chief contribution to the first was a clause in its
charter intended to assure that the remedy for a fall in the
exchange value of any currency would be sought not by the
route of depression and unemployment in that country, but
by the route of expanded imports and increased foreign investment on the part of the nation whose currency had risen by
comparison. This clause was general in its statement, but was
aimed directly at the danger of economic isolationism on the
part of the United States.

After the war Keynes explicitly repudiated those of his followers who had run his theories into the ground by advocating
perpetual and rigid controls of international trade, and isola-
tion of Britain from instability of the American economy. The
fate not only of Britain but of the world, Keynes deeply felt,
necessarily depended above all else on the capacity of the
United States to avoid depression.

The Role of Keynes

No Keynes was required to advocate abandonment of the
old standard, to create the general belief that unemployment
could not be cured by wage reductions, to advise governmental
deficits in depressions, or to support a low interest rate. All
these things had come into being before the *General Theory*
was published. It did not even take a Keynes to argue that
laissez faire did not work as it was supposed to, and that it
must be supplemented by governmental intervention. Many
economic writers before him had said as much, and tens of
millions of plain citizens believed it as a result of their own
experience.

A Keynes was, however, required to set at ease economists
in the classical tradition, plus a few generations of students

miseducated in their classes, who were confused by the diffe
ence between the seductively logical structure of classical do
trine and the spectacle of horrible reality before their eye
Keynes was capable of employing just as seductive a logi
embellished with the selfsame flourishes and with equally di
ficult mathematics, to explain why the system did *not* wor
He even included the standard calculus of marginalism ar
equilibrium.

Even more important, Keynes was able to prescribe rem
dies already obtained by the patient from the corner dru
store. He also promised a cure without serious discomfo
The choice was no longer between pure capitalism and pu
socialism. The theorists who recognized that something effe
tive had to be done heaved a sigh of relief after the assuran
from the most reputable possible authority that a major o
eration was not indicated. As an economic medical ma
Keynes brought scientific lore abreast of the intuitive co
clusions of the populace.

Keynes gave the members of the professional economic fr
ternity a new lease on life. They now had a pattern of thinki
which they could use in the positions of advice and respon
bility to which many were called in government, banking, ar
even business. They could go on with endless refinements ar
elaborations. No wonder that Keynes has converted the Briti
professional economists almost to a man, and that in t
United States his influence has swept almost all before
Adherence to laissez faire in the classical vein can rarely ne
be found except in the writings of members of the econom
"underworld," or among politicians or public-relations expe
defending some special interest against some special tax
regulation.

Quesnay had supplied ammunition for the agriculturis
Adam Smith had bolstered the industrial manufacturers; K
Marx had espoused revolutionary wage-earners; now J. I
Keynes had rescued the economic theorists. They desperate
needed a savior.

His major contribution was to adopt basic premises a lit
closer to existing reality than those of his predecessors. Havi
built this runway, he took off in high logical flight just as th
did, making little further contact with the ground. No da
existed by which to test whether intended savings did in in fa
vary from intended investment, or if they did, by how muc
Even if such figures had been available, it would have be
impossible to decide whether this difference alone was
sponsible for variations in income or employment that
curred, since so many other influences were at work at t

same time. Not only had Keynes little assurance from actual figures that the consumption function was as stable and predictable as he thought; he apparently did not even fully appreciate the importance of finding out. Experience in the United States since World War I indicates that he was wrong about it.

In view of the pretension of his theory to be better science and a better guide to policy than those of his predecessors, his neglect of careful corroboration from the world of nature is disturbing. He was, after all, a true follower of the classicists in his method of thought, if not in his conclusions. He relied on untested deduction from a few rather simple premises. He produced a new hypothesis—dazzling in its logical acrobatics, a useful tool of analysis, but still at bottom only an elaborate guess.

8

The Institutionalists—
Veblen, Commons, and Mitchell

When in 1819 the Italian Sismondi rejected both classical doctrines and classical method, not many followed his lead i
attempting to draw conclusions about economic principle
from the study of history. Yet a century later a few American
rediscovering the same trail, found that it seemed to be a pa
to the broad and refreshing prairies of a promised land.

Wealth, to Sismondi, was human well-being, not mere phys
cal riches. He distrusted Adam Smith's doctrine that men i
dividually seeking their own gain would necessarily bene
all. His skepticism arose not so much because he found fla
in the logic, but because the conclusion did not check wi
experience. Such checking indicated to him great unsolve
problems—poverty, unemployment, and recurrent busine
depressions. Sismondi declared that the necessary task w
to study human institutions and behavior in order to unde
stand and use economic science.

Early twentieth-century American economists, often call
Institutionalists, took just this direction. Sismondi might ha
read Thorstein Veblen with pleasure, if not with entire agre
ment; he would probably have collaborated eagerly with We
ley Clair Mitchell in his exploration of the business cycle.

Other scholars, mainly Germans, had tried to find their w
through the trackless economic jungle by way of Sismond
rough trail instead of via the elevated highway of the clas
cists. Traces of this endeavor are found in the thinking

Müller and List, briefly summarized in Chapter 5. These Germans were followed by others, usually grouped under the head of the Historical School.

The German Historical Economists

Adam Smith's tremendous influence was felt in Germany as well as elsewhere. But somehow in Germany his ideas could not be taken so seriously. The German state exercised wide control over economic life. In the universities economics was part of jurisprudence; candidates for the Civil Service studied it as future administrators, along with public law. Germans could not easily regard themselves as competing atoms afloat in a vacuum of laissez faire; they could more easily think of themselves as organic parts of a closely knit community.

About the middle of the nineteenth century German academic economists set out systematically to introduce historical ideas into economics, as Sismondi had urged. Wilhelm Roscher began by stimulating the application of theory to contemporary problems, a procedure which would necessitate adapting it to the current historical situation. Bruno Hildebrand went further; he wanted to transform the subject into a science of national development. Karl Knies made a frontal attack on the whole body of classical doctrine. He questioned that there were natural laws of human behavior—even laws of historical development. Theory, he believed, was relative solely to a succession of prevailing circumstances; it was an expression of opinion, not description of the world of nature.

Students taught by such men, who merely expressed attitudes, took them seriously enough to try to build a science on their views. In the 1870s Gustav Schmoller advocated gathering all sorts of material about actual human institutions past and present, and he set colleagues and students to doing it. When enough facts of this sort had been accumulated, he thought, generalizations would rise out of them by an inductive process. Until that distant time there could be no valid economic theory. The result of Schmoller's influence was a tremendous mass of descriptive monographic matter, but no reasoned doctrine of the sort that English and French thinkers recognized as economic theory.

Schmoller in his own works wrote of psychology—a matter, in his mind, largely of impulses or drives rather than of pleasure and pain in the Benthamite manner. He also reviewed technology and its progress, land utilization, the family, distribution of property and social classifications, various forms of government. In the end he expressed his own views (policy, colored fully as much by his conservative politic proclivities as by any objective studies. Schmoller's encycl pedic breadth enriched the economic landscape, but after h extensive travels in it he seemed to come out essentially th same man that went in. How was one to make any use fo scientific purposes, of so many disjointed factual observation:

Werner Sombart, born in 1863, attempted to illuminate th subject of economics by a gigantic history of capitalism. H did this from a socialist, though not orthodox Marxist, poi of view. His work does contain generalizations, including mar of dubious validity, especially those dealing with inhere capacities of races. The enormous mass of illuminating histori cal material which he put together seemed to guarantee litt more assurance of reliable scientific discovery by the proce of induction than came from the abstract deduction of th classicists.

Max Weber, born in 1864, on the whole had better succe in generalizing with the historical method than either Schmo ler or Sombart. Though he began as an economist, he is usual regarded as a sociologist, because of the breadth of his schola ship. Nevertheless he placed economics in its broad soci setting.

Weber's thesis that the rise of Protestantism made possib the triumph of the capitalist order in Europe, an observatio later brilliantly developed in England by R. H. Tawney, convincing. Other cultures have not yielded so large a sco to capitalist organization. Why was it, for example, that Chin with its rich resources and older civilization, was not t birthplace of the Industrial Revolution rather than Englan Weber found the causes of this fact largely in social forc with which economists had not customarily dealt.

The essence of a predominantly capitalist culture like th in Western Europe, Weber located not in rational self-dete mination of individuals, as the economic classicists taug but in a focusing of life about acquisitive and objective activ ties, which it is difficult for anyone to escape. Capitalism operated by an impersonal hierarchy in which the behavi of each participant is determined by his duties. Its type organization Weber calls bureaucracy, which he defines "a mechanism founded on discipline." This view of capital organization may be more relevant to what goes on than t doctrine which stresses individualistic competition for pro in free markets.

eblen, the Bad Boy of American Economics

Thorstein Veblen (1857-1929), an American living in the ame period as Schmoller, Sombart, and Weber, used much e same approach to economics. But he did not owe his in-ghts to them, and though in later life he became familiar ith their work he was just as critical of it as he was of that f most other economists.

The spirit of his books is that of the creative artist rather an of the ponderous German scholar. He read widely in istory, anthropology, psychology, and political science, but ough his great erudition enriched his writings it was not in e form of undigested lumps. When he made his style difficult nd used jaw-breaking words, as he often did, the reader could atch an ironic twinkle in his eye; he was showing that he ould mobilize jargon better than the stuffiest of academic uthorities—and with this heavy-treaded equipment could rind to powder their solemn banalities.

Each of his major books stated and embroidered a theme om his own creative imagination, and as a rule each theme as in essence a satirical comment on the culture in which he ved. His method was to describe some institution with the ost distant objectivity, as if he were a visiting anthropolo-ist from a higher civilization. The effect was more devastat-g than the angriest invective; no eloquence was required drive his points home. The loving care with which he elabo-ted his theses, however, gave evidence of something deeper an resentment toward the object of his analysis; he judged xisting society by reference to social standards that invoked s warm approval.

In a period when money-making was the ruling passion of rominent Americans, Veblen attacked the rich; he even did while on the faculty of a university subsidized by John . Rockefeller. In a civilization dominated by business he vinced profound disrespect for business enterprise. Recog-ized American economists had derived their theories from e British neo-classical school; Veblen made sly and learned n of the doctrines of his academic colleagues.

As a teacher Veblen offered no concessions to stupidity, ldom gave a grade above a C, and never lectured to under-raduates when he could help it. In college communities dur-g the Victorian age he seemed unable to avoid extramarital ve affairs, sometimes with adoring girl students. (For this e had some excuse in the immaturity of his wife.) In a

standardized middle-class society he cultivated a Van Dyk
beard and habitually wore sloppy clothing and a fur cap. Lik
most bad boys, he did not find favor with his superiors; h
could not long retain even the jobs which were offered hir
and never won high promotion. But also, like many a bad boy
Veblen was an original and fertile thinker—he derived les
from European tradition than any economist America ha
ever produced. And he was also the most solitary. He belonge
to no school or party.

Veblen's Life

Born on a Wisconsin farm of Norwegian stock, Veblen wa
one of a family of nine. Soon afterward the family moved t
Minnesota, where Veblen entered Carleton College at the rela
tively late age of twenty, and received his degree in three year:
His main interest was philosophy, in which he did postgradu
ate work at Johns Hopkins and Yale. In spite of his grea
ability he was unable to find a steady teaching job. For seve
years on the family farm he pursued his solitary education b
reading and thinking. Then he made another try and entere
Cornell as a graduate student, this time specializing in ecc
nomics.

One of his professors, the economist J. Laurence Laughli
was asked to accept a chair of economics at the new Universit
of Chicago and found a place there for Veblen as a graduat
fellow. At Chicago, Veblen remained for several years as
subordinate member of the faculty, produced much of h
basic work, and achieved a reputation. But the university wa
continually embarrassed by his eccentricities of opinion an
behavior and advised him to take a job offered by Stanfor
At Stanford he was socially just as much an outsider as :
Chicago, and he was soon forced to accept the only offer I
could get—from Missouri. This was an even more uncongeni
home for a man like Veblen, and Missouri proved for him th
end of the academic trail.

Veblen worked for a time in the Food Administration
Washington, became one of the editors of the refurbished an
liberalized *Dial Magazine*, and in the early 1920s was invit
to join the faculty of the newly established experimental Ne
School for Social Research in New York. There he lectur
to students attracted by his fame, but he spoke in such a lo
tone he could scarcely be heard in the back of the room, an
many who could hear him could not understand what he ha
to say. In 1929, the year when the regime of the financie
promoters, speculators, and big-business monopolists who

Veblen had satirized began its ignominious slide into the Great Depression, he died at the age of seventy-two.

Veblen's Principal Works

The theme of *The Theory of the Leisure Class* (1899) is that the social standards which determine behavior under Western capitalism are, beneath their modern trappings, much the same as those characteristic of barbarian societies. The money economy and the struggle for accumulation of wealth constitute new counters in the game, but the game exercises the same human traits.

The sign of high rank in both types of culture is "exemption from industrial toil." The barbarian rulers were warriors or priests; their status was won by predatory rather than productive exploits. Aristocratic virtues are the same today as then—"ferocity, self-seeking, clannishness, disingenuousness, and a free resort to force and fraud." Modern aristocrats, engaged in high finance and big business, exhibit these same "virtues," as do the retainers of financiers and big businessmen —bankers and lawyers.

The one distinction of the higher classes is that their activities are absolutely useless from the point of view of the humble citizen. The sign of success is lavish expenditure, which satisfies no real need but is a mark of prestige. Fine clothing in which one could not do any manual labor, a bejeweled wife, rich food, or useless learning constitute the "conspicuous waste" and "conspicuous consumption" which mark the man whom everyone wishes to emulate.

In *The Theory of Business Enterprise* (1904) Veblen paid his respects in a similar vein to the captains of industry—a term he invented. Industry using intricate technology is an efficient method of making goods that people need. But business enterprise is not the same as industry; on the contrary, it is a way of capturing control of some industrial process so that large sums of money can be squeezed out of it. Making money is very different from making goods; the two processes are often contradictory. The man who makes the most money is frequently the one who restricts production, eliminates competition, decreases efficiency, adulterates the product. Sometimes he does not deal in goods at all, but only in certificates of ownership, and devotes himself to stock-jobbing, market-cornering, mergers for financial profit. The activities of men in search of profit by these methods result in fraud on the consumer and small investor, panics, industrial depressions, and unemployment.

The concern for profit that motivates the captain of industry makes him fear overproduction of goods which, though they might be of great use to the consumer, would drive prices down below the level where profit would be maximized. In order to prevent such calamities, big businessmen habitually resort to "capitalistic sabotage." They lay off men or shut down their plants when prices fall too low, thus curbing "the inordinate productivity of the modern machine process."

Veblen loved to talk of big businessmen in terms of opprobrium usually reserved for organized labor. The restriction of production, which a pecuniary economy naturally led many captains of industry to organize, kept the workers poor; was it any wonder that laborers imitated their betters in order to find security or increase their inadequate wages?

In *Absentee Ownership and Business Enterprise in Recent Times* (1923) Veblen dubbed the closely knit financial control of big industry, which the corporate form of organization had made possible, the "One Big Union"—a term originally invented by the syndicalist Industrial Workers of the World (IWW) to describe their own goal. In *The Engineers and the Price System* (1921) he brought out his point by contrasting the function of the engineers and technologists—to increase production—with the function of the profit-seekers who hire them—to rig prices by interference with maximum production.

To Veblen the basic conflict in capitalism was not that between workers and capitalists in the Marxist sense, but that between the productive and the profit-seeking drives, each of them a major component of the capitalistic order. To him, the desire to produce, instead of being motivated by profit, was a natural human tendency, an "instinct of workmanship," as often as not thwarted by profit-seeking.

What Veblen Was After

Thorstein Veblen named few names, he cited few specific instances. He was not interested in bringing culprits to justice and suggested no specific reforms. If he had been an "agitator" his importance would have been less than it was. There were plenty of others calling attention to capitalist abuses in his day; abundant evidence with which to support charges of business and financial sins could be found in the reports of congressional committees and many other investigating agencies. But Veblen was after bigger game.

Veblen saw the schizophrenic chasm between the actual behavior of men under capitalism and the neat logical doctrine with which its defense had been rationalized. The doctrine, b holding minds in thrall, prevented both understanding and realistic action, by virtue of which modern technology might be permitted to confer the great benefits on humanity of which it was capable. Basic to everything else was to replace the accepted doctrine with concepts of another kind, concepts more relevant to the real situation.

In the first place, he thought, what was called economic "science" was out of date because it took no account of the evolutionary process. One state of society changed into another; there was no such thing as a fixed "system" about which eternal "laws" could be deduced. Karl Marx, to be sure, had written of social evolution, but his social theory was pre-Darwinian, just as his economics was in essence classical; society was not evolving by inexorable processes to a predetermined end.

Any given economic order must be understood in terms of the cultural patterns that predominate in it. People behave as they do because of the structure of the society and the ruling system of values. The study of how these things come into being, how they affect behavior, and what happens to the society as a result, is the truly scientific approach. Veblen was darkly amused by the survival in modern industrial society of barbaric values, which he saw expressed in the predatory activities of money-makers. He was equally intrigued by the fact that the rapid growth of machine production, which was the actual source of real wealth, called for entirely different values. The contrast between these incompatibles was the crucial problem of our society. How it was to be resolved he was not prepared to say. To him the important task for the scholar was to point out the contradiction in no uncertain terms.

Naturally, in his own specialty, Veblen rejected all the rigmarole of theory which had been built up about the "economic man"—an abstract atom in a free competitive market, a man motivated solely by the desire to gain pleasure and avoid pain. Psychology knew no such man, history and sociology did not know him, and he was not to be observed in reality. A good example of Veblen's style is the passage in which he ridicules this concept:

The hedonistic conception of man is that of a lightning calculator of pleasures and pains, who oscillates like a homogeneous globule of desire of happiness under the impulse of stimuli that shift him about the area. He has neither antecedent nor consequent. He is an isolated, definitive human datum, in stable equilibrium except for the buffets of the impinging forces that displace him in one direction or another. Self-imposed in elemental space, he spins symmetrically about his own spiritual axis until the parallelogram of forces bears down upon him, whereat he follows the line of the resultant. When the force of the impact is spent, he comes to rest, a self-contained globule of desire as before.[1]

Veblen's influence is of the kind which spreads, not through a wide popular audience, but through its impact on minds of men who in turn influence others. He had no celebrated disciples who preached and elaborated his doctrine as a whole; it never got into the accepted textbooks. But American economics—even American apologetics for big business—has never been quite the same since he wrote. Defenders of capitalism today emphasize technical achievements and great productivity rather than supposed automatic equilibrium of the system. Writers on economic subjects seldom fail to make at least a courteous bow to actual social institutions and values.

John R. Commons

A contemporary of Veblen, John R. Commons (1862-1945), likewise turned from theory to the study of institutions, but with quite a different temper and method of work. His father, a Quaker, was a newspaperman in Ohio, his mother a Presbyterian and a graduate of Oberlin. John himself, partly through the aid of his mother, worked his way through Oberlin. In doing so he held jobs in printing shops and joined the union. He also read Henry George's *Progress and Poverty*.

As a result of his struggle for an education and his experience of the difference between the theory and the practice of economics, he decided to investigate the subject for himself. This he did by taking graduate work at Johns Hopkins under Richard T. Ely, where he spent almost as much time having a look at what was going on in the neighborhood as at his desk. Ely, familiar with the German historical economists, believed in the inductive method, and Commons put it into practice.

A job at Wesleyan he held only a year, because the administration did not like his practice of discussing with his students the problems of the day and the neighborhood instead of the theory in the textbooks. Then he went to Indiana University, became interested in reforms and such unorthodox matters as the cooperative movement and municipal ownership of utilities.

The idea that economics was merely a part of the broader subject of sociology was spreading, and Commons was invited

[1] *The Place of Science in Modern Civilization* (New York: B. W. Huebsch, 1919), pp. 73-74.

to fill the new chair of sociology at Syracuse. There he wrote a series of articles under the title *A Sociological View of Sovereignty*, in which he set forth the thesis that the owners of private property held the chief power in our society, and that as they used that power to encroach on the rights and welfare of others, the citizens turned to the state to curb the abuses. Hence governmental sovereignty grew along with the power of monopoly. The men of wealth who contributed to Syracuse confirmed his idea of their power when, alarmed by his teachings, they withdrew their support and his chair was abolished.

For several years Commons was allowed to practice his talents for investigating and dealing with actual economic institutions by working with the United States Industrial Commission, which was looking into strikes, labor organizations, and employment practices. From there he went to the National Civic Federation, an organization formed jointly by employers and national labor leaders to promote industrial peace.

In 1904 Commons was invited back into academic life at the University of Wisconsin. Under the progressive administration of Governor Robert M. La Follette (the elder), the university was called upon to help in framing a code of social legislation, and this task gave Commons his chance to participate in economic action while he was teaching. Labor legislation, taxation, public utility regulation, and civil service felt the impress of his expert advice.

His major accomplishment, however, which he carried out with aid of a number of brilliant and devoted students, was to write the history of American industrialism, with particular emphasis on labor organization. *A Documentary History of American Industrial Society* (1910) in eleven volumes, was followed by the work for which he is chiefly known, *History of Labor in the United States*, first published in 1918 and later supplemented by additional volumes written by his faithful co-workers.

Nobody can become familiar with the actual history of labor, of its political and industrial struggles, and of the development of collective bargaining, without being impressed by the fact that the labor movement is, in our culture, just as natural and inevitable an accompaniment of the Industrial Revolution as the modern corporation. Abstract theories about wages, or about the rights of labor to organize, tend to dissolve in the turbulent stream of experience. By his scholarly documentation of history Commons exerted as much influence on American economic opinion and practice as if he had evolved a striking new abstract generalization in the realm of theory. And his pioneer work opened up a realm of knowledge concerning which neither the doctrines of the classical economists nor the revolutionary concepts of Marx seemed to be very illuminating.

Commons did not ignore other important institutional influences on our society—for example, banking and money, with their inflationary or deflationary effects. Nor did he lose sight of the aim of drawing theoretical conclusions from his studies. These he attempted to sum up in a final contribution, *Institutional Economics* (1936).

We have a society, he pointed out, in which the cooperation of everybody is required to gain the maximum benefit, but in which people are continually quarreling about possession of property and division of the product. Collective controls have been developed in the general interest to routinize and discipline the particular conflicts of interest. These controls govern bargaining—the subject mainly dealt with in classical studies of markets and prices—and also two other institutions prevalent in our society, managerial transactions and rationing. A managerial transaction is one in which a superior exercises his power to create wealth, and a rationing transaction is one in which a superior makes decisions about distribution of the benefits or responsibilities arising in the process of wealth creation.

To Commons it was an open question whether in the future evolution of our society most of the important decisions would be made by bargains among equals, as it is supposed to be in democratic capitalism, or by managerial decisions, as it is supposed to be in dictatorships like those of communism or fascism. His own bias lay on the side of democracy, and so he believed in maintaining or creating equality of bargaining power. In our culture, of course, equality of bargaining power could not rest solely on individual action, but must often be sought by collective means, as in the case of trade unions, cooperative consumer societies, farmers' organizations. What his philosophy in essence looked forward to was a harmonious balance among great interest groups, with a wide distribution of ownership among the people. In all this, democratically controlled government must, he thought, play an important role, as must the law as interpreted by the courts.

Wesley Mitchell and His New Beginning

Wesley Clair Mitchell (1874-1948) devoted his life to an attempt to lay the foundation of a genuine economic science which might be as reliable and useful as any natural science. In order to do this, he believed, it was necessary to study, not nature in the romantic sense of a "natural" primitive society, or in the mystical sense of a "natural" law which man can disobey, but nature in the sense of the processes which actually may be observed in economic life. It should be possible to do this just as processes of nature are observed in the realms of physics, chemistry, astronomy, geology, or the biology of insects, plants, and mammals. If economics were to be a body of verified knowledge, he was fond of saying, it must rest upon a careful study of human behavior, not merely on hypotheses deduced from a few simple premises or isolated observations.

True scientist that he was, Mitchell learned to approach his work by discounting preconceptions, so that he could better observe and interpret what actually happened. This task was not easy in economics, since through the centuries so large and varied a body of doctrine had been developed. Mitchell did not lack interest in these doctrines—one of his most popular courses at Columbia was a brilliant series of lectures on the history of economic thought. Nor did he spend much effort in undermining or opposing them. He simply waited until the work he regarded as important might reveal how valid any of the bodies of theory might be. Some, he expected, would be supported; some would be disproved. A good deal of the existing doctrine, however, he came to believe, would turn out to be irrelevant in a scientific sense, since the hypotheses were so stated as to be incapable of either proof or refutation by reference to observable facts.

How was human behavior in economic activities to be systematically observed? Lack of a reliable technique had hampered most previous students of institutions in arriving at anything like precise results. They had employed description and used words, but after their best efforts there might be much of the subjective left in their conclusions. Their generalizations were, after all, opinions.

Mitchell saw that the essence of a truly scientific observation was measurement. One had to know what was happening not merely in general terms but in terms of quantity. From the beginning he called to the aid of his analysis the art of interpreting statistical information. Measurement of changes in prices, in production, and in many other factors significant to economic problems was, he saw, in essence measurement of the economic behavior of society, in whole or in part. It was, too, a sort of history, since all that one could measure in this way had already occurred.

Emphasis on the use of statistics as a tool of investigation is now widespread, but Mitchell was a pioneer in understanding the importance of the technique. Many had employed selected figures to support conclusions arrived at by other means; Mitchell conscientiously used all the relevant figures he could find as a means of discovery and resolutely refrained from formulating conclusions unless the figures warranted them. Fortunately he came at a time when economic statistics were about to become abundant because of the necessity for records in public administration and business practice.

Thus Wesley Mitchell made a really new beginning in economic research. It was a task that would perhaps never be finished, and it would have to depend on the talents of many skilled investigators. Theory might be slow in coming but once conclusions became possible, they would be solidly based. Mitchell, recognized during the latter part of his life as the most eminent American economist, won his reputation not as the author of brilliant improvisations or neat processes of logic which might or might not be relevant to economic life, but as the leading founder of a type of economic learning which could grow by accretions and which could enlist the contributions of as many in the economic fraternity as could master its disciplines. This kind of cooperation characterizes all sciences worthy of the name.

Mitchell's Life and Work

Wesley Mitchell was born in Illinois, the eldest son of a former Army doctor in the Civil War who went into farming. The family was a large one, and much of the responsibility of managing the fruit farm fell upon young Wesley at an early age. This contact with hard reality left its mark upon him. But he also learned the tricky possibilities of logic, during theological arguments with a great-aunt who was a devout Baptist. He delighted in upsetting her positions by adducing evidence not comprised in her premises.

At the young University of Chicago, Mitchell was profoundly influenced by an unusually distinguished faculty. His professor of economics, J. Laurence Laughlin, was a monetary theorist who fortunately set Mitchell to work on a study of the Civil War greenback episode for his doctoral dissertation. But Laughlin was of the classical tradition and in spirit had much the same effect on Mitchell as his Baptist great-aunt. Veblen's point of view Mitchell found much more sympathetic, and he accepted Veblen's criticism of the classical method and premises. He always felt, however, that Veblen's own conclusions lacked solid support. They were plausible generalizations—but where was the marshaling of facts that would show whether, or to what extent, they were true?

John Dewey gave him new insights into psychology, showing that any doctrine which assumed that human behavior was based on rational calculation was likely to be unsound. Dewey as a philosopher confirmed Mitchell's belief in the importance of what actually happened as a result of what men actually did, as distinguished from men's rationalizations of their behavior. He began to try to explain the mental processes of the orthodox economic theorists rather than taking their theories seriously. Jacques Loeb, the distinguished biologist, taught him the elements of scientific method.

In his doctoral dissertation, *History of the Greenbacks*, Mitchell began to use the methods that were later to characterize his work. His analysis involved figures concerning the amount of money poured into the system by the Civil War issues of fiat money, and figures on the trend of prices in general and in detail. The comparative statistics in this and a subsequent work, *Gold, Prices and Wages under the Greenback Standard*, did not support the quantity theory of money in any exact sense of a mathematical relationship between money and prices, but allowed plenty of room for psychological influences—what would now be called expectations—on the part of the people. He furthermore broke new ground in showing that inflation and deflation had different effects on different kinds of prices and levels of income. Most economists had talked of this type of process much as if it had a unified impact throughout the economy; many still do. Mitchell in his first major work learned the importance of breaking up large aggregates into smaller components and tracing the processes of change as they make their way through the economy.

After his graduation Mitchell taught for a while at Chicago and then went to the University of California. There he began monumental statistical study, *Business Cycles* (1913). The book summarized the few cycle theories of other economists who had gone before him, identified the concept of the business cycle itself as a continually recurring phenomenon of any business order, divided it into characteristic phases, and showed, so far as the figures were available, what the history of the cycle in the United States had been. The large volume made his reputation, but it was characteristic of Mitchell that he regarded the book merely as a beginning, returned to the subject whenever he had the opportunity, spent the latter part of his life on it, and left his work unfinished when he died.

Mitchell now shifted to Columbia, where he could be closer the centers of economic activity. During World War I he was called to serve the government as one of the expert ad-

visers on mobilization of the economy and control of prices
This task involved extensive use of existing statistics and th
beginning of a large number of new series. Rich and expand
ing material existed for the kind of research to which Mitchel
was dedicated. As soon as possible after the close of the war
he organized, with a few associates who held similar views o
the need for scientific method in economics, the Nationa
Bureau of Economic Research. He became its Director of Re
search, and in this connection did most of his work for th
remaining twenty-eight years of his life.

The National Bureau grew from small beginnings to a larg
institution with many economic scientists on its staff and
long list of publications. It is not a commercial institution bu
is devoted entirely to basic research, supported by contribu
tions from philanthropic foundations and many others. Ab
sence of bias in its reports is guaranteed by a board of direc
tors representing many varieties of opinion and background
The Bureau is also linked with prominent universities.

Work of the National Bureau under Mitchell

From the time of the foundation of the National Burea
until Wesley Mitchell's death, it is difficult to draw distinc
tions between his own work and that of the staff of associate
who worked with him. His name is not signed to many c
the volumes, each of which is attributed to a specific autho
or authors. He was the most generous of collaborators an
regarded the whole enterprise as a cooperative one, in whic
each scholar, after the benefit of consultation with his fellow
was entitled to credit and responsibility for his own contribu
tion. Yet Mitchell's spirit pervaded the whole, and he wa
chiefly instrumental in mapping out the program and supe
vising its execution.

Two main subjects, closely related to each other, const
tuted the foci of most of the Bureau's work. One was th
national income, the other, the business cycle. Together, thes
subjects stand at the heart of the behavior of the econom
for one of them measures the total annual product of th
economy over a long series of years, together with its origi
and its distribution among the various kinds and classes
income receivers, and the other measures the fluctuations
economic activity as a whole and in detail, illuminating th
processes by which these fluctuations occur and their effect

Estimation of the national income, its components and i
distribution, has now become a highly standardized proces
and the results are widely published and used for man

purposes. The task, in which the National Bureau was the chief pioneer, has been taken over as a regular function of the United States Department of Commerce. Such questions as how much is produced by agriculture, various manufacturing industries, mines, and construction activities, what we pay for the services of transportation, wholesale and retail trade, government and the professions, how much profit business enterprises make and how much they invest, what consumers spend and what they save, are known within a narrow margin of error (except in the case of consumers' savings) for many years past and for current years soon after the event. With voluminous information of this kind it is possible to talk about the actual economic order rather than about an imaginary one deduced in the abstract. This constitutes a veritable revolution in economic thought.

The work of the National Bureau in the business cycle consists mainly of a large number of monographs on various aspects and parts of the phenomenon. It was Mitchell's plan to bring together all this material and fit it together in a definitive description of the cyclical process. But the scope of the necessary investigation expanded as the research proceeded, and unfortunately Mitchell did not live long enough to complete his life work, though he survived beyond the Biblically allotted threescore years and ten. What he did do was to publish a preliminary volume in 1927, *Business Cycles: The Problem and Its Setting,* to write, with Arthur F. Burns (his successor as Director of Research of the Bureau), a massive technical work, *Measuring Business Cycles* (1946), on the methods of measurement employed in the research, and to make a "progress report"—published posthumously in 1951, thirty-eight years after his first study of the subject. This last volume, *What Happens During Business Cycles,* he hoped to finish before death overtook him, but he did not quite succeed in doing so; Arthur F. Burns edited it and summarized it in an introduction. Mitchell's patient persistence and his tender scientific conscience are revealed by the statement in the first chapter of his last book: "Even now what we can say is ill proportioned, tentative, and subject to change as the investigation proceeds."

What Kind of Theory?

What sort of conclusions did Mitchell draw from these years of research into the vast array of figures arranged in time series? Not a theory about what "causes" depressions, not, for example, an abstraction like that of Keynes concerning the consequences of a disparity between saving and investment. The only kind of abstraction Mitchell made arose from observation of a large assortment of facts; for example, he attempted to show what a "typical" cycle is like, as distinguished from a large number of actual historical cycles none of which is exactly like any other. And he described the typical cycle, as his title suggests, by recounting "what happens" during it, as a rule. His figures led to generalizations such as those summarized by Burns—"that crop production moves rather independently of business cycles, . . . that production typically fluctuates over a much wider range than prices, that the liabilities of business failures usually turn down months before economic recovery becomes general and turn up months before recession . . . . that orders for investment goods tend to lead the tides in aggregate activity," etc.

A cycle is revealed as highly complex. All series do not turn up or down at the same time. "Every month some activities reach cyclical peaks and others decline to their troughs; so that expansion and contraction run side by side all the time. But the peaks tend to come in bunches and likewise the troughs." It is by analyzing the differences in timing and the differences in amplitude, Mitchell wrote, that one may see "how an economic system of interrelated parts develops internal stresses during expansions, stresses that bring on recessions, and how the uneven contractions in its various parts pave the way for revivals."

Critics sometimes urge that this type of theory, like former attempts to generalize on the basis of induction, is unsatisfactory. They say that it is descriptive merely and does not lead to the logical analysis which would make possible the discovery and elimination of a "root cause." To be sure, it may show how disturbances spread throughout the system and lead to depression, how turning points occur, and how revival gains momentum. But what is the microbe that brings on the disease?

This attitude reveals a fundamental difference between Mitchell and the adherents of the traditional deductive method. They tacitly assume, much as the classicists did, that there is a normal state of equilibrium at which the economic order would remain if some disturbing factor could be eliminated. Even Keynes' theory was stated in terms of departure from equilibrium, though he acknowledged that equilibrium might be reached at a level of activity too low to permit full employment. But this is one of the assumptions that Mitchell abandoned. He posited no "normal" or "equilibrium" state of the economy; and he found none in its actual behavior. Busness and employment have been continually in flux upward or downward. The cycle is itself normal in any business economy that we know, according to Mitchell's view. Causation takes on a different meaning in these circumstances. It inheres in the complex interaction of many factors, not in a single "cause," unless one thinks of the cause as the whole make-up of the economy. Mitchell's abstraction from the multitudinous facts of reality is the "typical" cycle, not a state of equilibrium.

Was Mitchell then hopeless of moderating the swings of activity that engender so much distress? By no means. Since the process itself is the result of the way people behave, it can be modified if they modify their behavior. Scientific discovery has been proved useful in enabling men to achieve results that seem to them worth while; verified knowledge may easily reveal points at which carefully planned intervention can bring desired results in any living process. No system of human institutions is fixed for all time; our own economy has undergone extensive changes. Once the cycle is thoroughly understood, further changes may be recommended to rob it of most of its terrors.

Mitchell differed from those who had prescribed remedies for depression and inflation chiefly in wanting to be very sure of knowing what he was talking about before writing his prescription. He was not willing to move from the laboratory to the testing ground on the basis of a guess, however plausible. Even if his empirical method should never reveal what he was looking for, he knew of no other on which he could rest his faith. The methods of modern science have already made possible many apparent miracles; in the long run they may be able to yield equally startling results in the analysis and molding of human behavior.

9

The Use of Economic Ideas

Any reader of the great economists is likely to be impressed
by each doctrine as it passes in review. Each prominent body
of economic theory bears the aesthetic impress of masters of
thinking; there is no more absorbing logic to be found in
human lore, unless it be in the closely related exercises of
metaphysics or mathematics. Those susceptible to the charms
of the Goddess of Reason are likely to become in turn Physiocrats, disciples of Adam Smith, Malthusians, utopian socialists, Marxists, Marginalists, Keynesians. Yet one cannot con-
sistently adhere to all these doctrines at the same time, since
many of their tenets are mutually contradictory.

Some devotees apparently choose one or another variety of
theory as if choice were a matter of taste. Should one furnish
the rooms of his intellectual life in the style of the early American, the Victorian, the "modern," or one of the other varieties
or sub-varieties of décor? There is no reason why anyone need
hesitate to do so if it is clear to him that all he is doing is pursuing an adventure in experience, to satisfy a personal taste.
Unfortunately, however, many adherents of doctrine mistake
systems of logic for ultimate and universal truth. They insist
that others must obey the principles which attract them. They
mistake a neat design of ideas at their intellectual window
for the outdoor world. No one can reckon the human misery
caused by those who cannot see beyond the curtains of their
ideas.

Can Economics Be a Useful Science?

The economic theories of the past, we have seen, have had
their uses. The great ideas have been in some degree at least

relevant to their times; they have transmitted light, filtered through the climates of their periods. They have revealed, though often implicitly only, the social pressures and needs of contemporary persons. They have been employed as rationalizations and defenses of armies on the march, and of armies resisting change.

Each great body of economic doctrine has not only been useful in the humble sense of service to interests, but has also carried the banner of science. That is, it has pretended to be an aid in the accurate and systematic understanding of processes in the world of nature. Like physics, chemistry, and all the other natural sciences, it has attempted to explain the network of actual relationships in such a way that men can predict the consequences of what happens or is made to happen. Understanding of this sort enables men to avoid dangers and gain what they desire. But, in their scientific pretensions, economic doctrines have not been so successful as in their logic-spinning. It is one thing to be able to say, given certain premises, that certain conclusions must follow; quite another to be sure that the chosen premises are typical of the objective world, or that what is excluded from any particular logical process may not be more influential in determining the future than what is included.

At least two major circumstances make it more difficult to reach usable conclusions in sciences dealing with human behavior than in sciences dealing with the behavior of matter and energy not organized in the forms characteristic of what we call living beings.

First, given a small sample of human behavior, one can be less certain that this sample will be typical of the behavior of men in general than that, given a small sample of the behavior of inanimate matter or energy, the sample will be typical of the behavior of all other instances of the same kind. The reaction of a few milligrams of chemicals in a laboratory test tube is likely to be repeated when chemicals of the same nature and purity are combined in the same proportions in any quantity or location. If this likelihood were not overwhelming, chemistry would not be a very useful science. Yet individual men combined in societies of any magnitude may react very differently in different times and places. This fact is what makes it precarious to build intricate logical systems of economic thought on rough observations of a few simple instances of human behavior. The risk of error is compounded when, as often has happened in economic thinking, the specific instances chosen are not even observed but are imagined. This is the case when men are presumed to be completely rational and to be motivated solely by desire to gain wealth or avoid loss.

An even greater difficulty in gaining systematic and reliable knowledge of human behavior is the fact that men themselves continually change their patterns of action and thinking. New ways of acting produce new situations which in turn stimulate more change. Any social science worthy of the name must be a science of becoming, not merely a science of what would happen if conditions were fixed. Perfect knowledge of what happened in 1950 and why, provided such knowledge were available, would be of some service in understanding what is likely to happen in 1960, but when 1960 comes what we shall need most of all is relevant knowledge of what has just happened and what *is happening*, if we are to use what we know in guiding next steps.

The question is whether we can have an economic science capable of describing the processes of social behavior accurately enough so that it will be of use in making the decisions which will affect the future. What is needed is the information, interpretation of which will enable us to make useful plans.

The Argument about Economic Planning

Economists have always looked with approval on the fact that individuals and business concerns make plans and try to carry them out, but not all of them have approved planning on a national or international scale. Nations did plan extensively in the age of mercantilism, and though some of the planning was necessary, much of it sought empty goals and led to undesirable results. Classical economists contended that people should be left almost free of control in their economic activities.

In the present world, governments have grown steadily more important as forces in the economy. Even if governments never exercised any direct powers over production, prices, and trade, their own policies in taxing, borrowing, and spending, and the effects on banking and money of what they do, would be certain to bring large consequences to the national economy as a whole. In addition, governments have become producers of many goods and services, not only in socialist countries but in most others, and have exercised wide regulatory powers over economic practices.

In nations like the Soviet Union, which have abolished private enterprise on principle, the government as the sole o ganizer of production and distribution was compelled to install a planning system in order to make any sense out of its economic activities. In most other nations, even those under governments called socialist, there are large sectors of the economy still in private hands. The increase of governmental activities in these countries was brought about, not mainly in consequence of belief in socialist doctrine by the citizens, but because of practical judgments as to what the community needed in specific instances, and the conclusion that governmental agencies were better fitted to supply these needs than private organizations seeking profit, if indeed private agencies could supply them at all. Even in non-socialist countries the idea that coordinated policies adjusted to the interest of the whole economy should be formulated and followed was certain to make headway. Naturally enough this kind of advance calculation is often called planning, though it is not planning in the sense in which the word has come to be applied to the regime in the Soviet Union, where every decision is supposed to be dependent on a central dictatorship.

The drift toward the choice of goals for a national economy as a whole, and the formulation of policies designed to achieve these goals, has been bitterly attacked by a few theoretical economists who still adhere strongly to the tradition of classical economics. Prominent among these are two who carry on the ideas of the Austrian school of marginal analysis—Ludwig von Mises and Friedrich Hayek. The main burden of their argument is well stated in Hayek's *The Road to Serfdom*.

In an economy of competitive private enterprise where free markets prevail, argues Hayek, individual businessmen seeking their own profit will, without even trying to serve the general interest, naturally make the decisions which give the consumers what they want most at the lowest possible prices. Both producers and consumers will be free. But any higher authority seeking to substitute its judgment for theirs will not have such good criteria for making its decisions, since it will be deprived of the signposts constituted by the prices and profits of a free market. It will make mistakes. Any mistake made by a governmental authority is bound to be more serious than one made by an individual businessman, since the small competitor injures mainly himself by a wrong judgment, whereas a national authority injures everybody. In order to correct its mistakes and bolster its prestige the government will be forced into more and more intervention, continually extending its power until we are all slaves. According to this theory, the oppressive Russian dictatorship is a consequence of the Communist decision to plan. And unless other peoples base their economies on the free market, automatically regulated by its own processes, they are on the road to ruin and are bound in the end to be subject to dictatorships run by scoundrels like the Nazis or the Bolsheviks.

Whatever anyone may think about it, nations have been compelled to undertake a good deal of planning because of stern necessity. No modern nation can run a war by giving free rein to the profit system and consumers' choices, because consumers do not individually buy aircraft carriers, armored divisions, or atomic weapons. The United States, among other belligerents, planned extensively even in World War I. And when, not many years afterward, it was plunged into a profound and stubborn depression, again the people turned to government to do many things that the incentive of private profit did not lead businessmen to do. The depression was shortly succeeded by another war, and that by a renewed need, not only for national defense, but for concerted action on an international scale.

Governmental activities of the present magnitude must be planned. If Messrs. Hayek and von Mises can show us the way to pure laissez faire it is their privilege to do so, but merely to contend that it would be better than the existing scheme of things does not seem to get us anywhere. And it looks like a doctrine of despair to say that because government must exercise some control we are all bound for a dictatorial Gehenna. It would be more pertinent to inquire how, so far as planning is necessary, mistakes may be avoided and how bad planning may be made better. And it would be more realistic to inquire in what types of economic activity free markets are possible and more useful than planned initiative.

No area in which private enterprise is now dominant is likely to fall a prey to socialism unless it fails to perform reasonably well. A mixed economy may be more satisfactory than either one in which everything is socialized or one in which nothing is socialized. The doctrinaire arguments both for and against socialism seem more and more sterile as the years pass. But there is no good reason why, as a democratic national community, we should not try to apply informed foresight to the achievement of the common goals that seem important to us. And there is no reason why we should not try to achieve these goals by a combination of action by government, business executives, farm and labor organizations, and individuals. In cases where goals cannot be agreed upon, we shall have to decide the outcome by our usual confused democratic processes.

Economics as an Aid to Planning

If we lived in the kind of world pictured as desirable by the classical economists, economic science would be of little use, except perhaps to tell us that such a world is desirable. For in that world native shrewdness alone would, according to classical theory, serve to seek out the best opportunities for profit, and nothing more than this would be necessary. But in a world where decisions must be made involving the welfare of the whole community or large segments of it, reliable information and pertinent analysis are essential. How is such information obtained?

Statistical method offers a means of avoiding the errors to which generalization on the basis of too small a sample of human behavior is prone. One man's action differs from that of another, but the behavior of a large aggregate of individuals over a sufficiently long period of time often reveals a greater uniformity. Statistical method is a way of discovering the behavior of aggregates, and so testing the conformity of general behavior with that of samples. An amazing amount of reliable information about customary economic behavior has been derived from the immense volume of statistics that have become available in recent years. Much more remains to be discovered, but at least some reliable methods of discovery have been invented and tested.

Statistical information, however, is necessarily information about the past. In a rapidly developing society, how shall we be prepared to deal with the process of change? There can be no certainty about the future of any human experience, but the broader the knowledge and the deeper the analysis of what has happened, the better is the chance of guiding present action to achieve future goals. It is now possible, as it was not in former centuries, to lay down quantitative tests for the success of many chosen policies. Do we want to control the volume of credit or the amount of money in the hands of the public? Do we wish to stabilize prices? Is it desirable to increase total production, or to direct productive forces toward one type of output rather than another? Do we aim to minimize unemployment? Figures are now available, some of them month by month or even week by week, to test success in these or hundreds of other possible objectives. If one method does not work well it may be improved, or abandoned for another. By a series of approximations the validity of conclusions about the effects of deliberate changes in existing practices may be judged. All this is far from the methods of exact laboratory science, but compared with it, past ways of deciding public policy were like steering a car in the dark without headlights.

If economists had the time to prepare a careful inventory of all the reliable information now available about the economic order that was unknown fifty years ago, the comparison would be startling. The spinners of theoretical webs now have much more than insubstantial gossamer with which to work. More and more it becomes possible to check opinions against experience.

There is still much to be done. Indeed, as in other sciences, every discovery opens up a wider area of unknowns. Yet there is ground for believing that at last there is beginning to be an accumulation of tested conclusions that cannot be blown away by the next wind of doctrine. The task of national and international housekeeping, with its critical need for guidance, may yet come to rest on verified knowledge as much as on blind habit or prejudice.

Index

157